INFORMATION MECHANICS

CONSTRUCTING THE COSMOS

ROBERT WILES

MB BS FRCGP FRACGP FACRRM

INFORMATION PRESS

Don't read this book!
Why? Is it poorly written? ......No.
Is it bad science? ......No.
Is it slow and boring? ......No.
Is it hard to read? ......Yes!
It will take some effort to read, assimilate and gain understanding.

So, why should I not read it?

It will turn your world upside down and then shake it around. This book challenges all aspects of accepted science and will challenge your world view even if you don't understand your own world view (yet). You will be forced to answer some of life's Big Questions about the nature of reality. Wiles pulls the rug out from under some of the preconceptions that often underpin Science and Philosophy.

This book is about science of all kinds and areas, but brings implications, not only for how Science is analyzed, but also Philosophy, in the area of worldviews and the assumptions of life that we all have. Please read on and enjoy the ride!

Dr Ken Meyer PhD

Electrical engineer - former Lecturer in Electronics
Canberra Institute of Technology

Dr Robert Wiles is not one to look at things superficially. In his new book, he delves into the very fabric of our universe, not simply of its structure, but of the paradigm which we use to understand and interpret it. He seeks to bring a new language to this understanding of what makes the world and indeed the universe go round, or rather, oscillate!

Dr David Learoyd FACRRM

Rural Generalist Physician

When I first started studying,
I saw the world as composed of particles.
Looking more deeply I discovered waves.
Now after a lifetime of study,
it appears that all existence is the expression of information.

JOHN WHEELER[1]

TO

My parents and teachers
who taught me to question

First Published in 2026
By Information Press (Australasia)
P. O. Box 159
Cooma. NSW. 2630
Australia

Website: robertwiles.com.au

ISBN-13: 978-0-9875622-3-4

National Library of Australia
Cataloguing-in-Publication data is available from:
http://catalogue.nla.gov.au

Book and Cover Design: John Locklair, Blue South Creative, South Carolina.

Printed by: Lightning Source
Cover photo: Adobe Stock

Artificial Intelligence (AI) was not used intentionally or knowingly in writing this book. The views expressed are those of the author alone and should not be purported to represent those of any other person or organisation, except as specifically referenced.

INFORMATION MECHANICS

Foreword 11
Preface 13

1 INFORMATION MECHANICS: WHY WE NEED IT
1.1 Information throughout History 18
1.2 Recognizing the significance of Information 20
1.3 Understanding the current Materialist Paradigm 21
1.4 The Changing Zeitgeist 22
1.5 Information Mechanics 23
1.6 Introducing Mind 24

2 EXPLORING INFORMATION
2.1 Re-envisioning the Universe 27
2.2 Enlisting the Tools of Science 28
2.3 The Fundamental Entities in the Universe 29
2.4 Phantom Particles 30
2.5 What is Information? 31
2.6 The Significance of Mind 33

3 THE CHARACTERISTICS OF INFORMATION
3.1 The Infosphere: Hardware, Software & Information 38
3.2 The Function of Information 39
3.3 Meaningful Information 40
3.4 Verifying Meaning 41
3.5 Information and Time 42
3.6 The Importance of the Quantum World 42

4 SURPRIZING OUTCOMES
4.1 Our Mathematical Universe 45
4.2 A Quantized Cosmos 46
4.3 The Universe is finely tuned for Life 47
4.4 The Evolution of Science 48
4.5 Immateriality 49
4.6 Information Generation 50
4.7 Conservation of Information 51
4.8 Information as the Fundamental Entity 51

5 **BUILDING COMPLEXITY**

5.1 What is Life? 55
5.2 Non-life into Life 56
5.3 Information in Life 58
5.4 Systems Biology 59
5.5 The Origin of Life 60
5.6 Self-organising Systems 61
5.7 Invoking an Algorithmic Origin 63

6 **THE MIND CONUNDRUM**

6.1 Understanding the Human Mind 65
6.2 Self-awareness 66
6.3 Theory of Mind 67
6.4 The Quantum Mind 68
6.5 Models of the Mind 69
6.6 Quantum Biology 70
6.7 The Resonant Mind 71
6.8 Mind Fields 72
6.9 Sleep 74
6.10 Shifting our Thinking 74
6.11 The Mind-Brain Complex 76
6.12 Monism or Dualism 77
6.13 Interpretation of Meaning 78

7 **CLASSIFYING INFORMATION**

7.1 Different descriptions of Information 79
7.2 Space-time 80
7.3 Cyberspace 81
7.4 Mindspace 82
7.5 The Quantum Domain 84
7.6 Dimensions 87

8 **RECONCILING THE CLASSICAL AND QUANTUM REALMS**

8.1 Mathematical Representation 89
8.2 Black Holes 90
8.3 White Holes 91
8.4 Universe or Multiverse 92
8.5 Reconciling Space-time with the Quantum Domain 93

8.6 Quantum Leakage ... 94
8.7 Emotions ... 96
8.8 Our Holographic Universe ... 96
8.9 Needing to Dig Deeper ... 98
8.10 Probabilities and the Quantum Realm ... 99
8.11 Finding the Electron ... 102
8.12 Plasma ... 104

9 CONSTRUCTING THE COSMOS FROM INFORMATION

9.1 A Summary of our current Paradigm ... 107
9.2 Information Gaps and Discontinuities ... 108
9.3 Front-loading the Universe ... 110
9.4 Evolution of the Gaps ... 111
9.5 Misnomers ... 112
9.6 Punctuated Equilibrium ... 113
9.7 Information Generation and its Origins ... 113
9.8 Reviewing the Evidence ... 114
9.9 Constructing the Cosmos from Algebra ... 117

10 INFORMATION MECHANICS: A NEW UNDERSTANDING

10.1 Counterflow of Complexity ... 119
10.2 Scientific Reductionism versus the Integration of Information ... 120
10.3 Mechanism versus Agency ... 121
10.4 Top-down Thinking ... 123
10.5 Testing the Information Model of the Universe ... 126
10.6 Shifting to an Information Paradigm ... 126
10.7 Conclusion ... 129

Acknowledgements ... 133
Notes ... 135
Index ... 149
About the Author ... 155

1. Cited in Gerald Schroeder, "Can God Be brought into the Equation?" review of *Science and Religion: Are They Compatible?* ed. Paul Kurtz and Barry Karr, in the Jerusalem Post, May 23, 2003, 13-B; See Pearcey, N: Total Truth: Liberating Christianity from Its Cultural Captivity. Crossway, Wheaton, Il (2008) p.179.

FOREWORD

Science, in its purest form, is the reflection of God's genius. With each passing generation, new discoveries draw back the curtain on His universe, revealing more wisdom and beauty that have always been there, waiting to be seen.

In *Information Mechanics,* Dr Robert Wiles invites us to behold the wonders of creation through a lens both scientific and sacred. He takes the staggering complexity of the cosmos – with all its grandeur, its order, and its secrets long concealed – and presents them in a way that stirs both the mind and the heart. His purpose is not merely to explain what can be measured, but to illuminate what can be marvelled at: the greatness of God.

This is not an easy read, nor should it be. It is a necessary one. For through these pages, we are reminded that every law of physics, every unseen particle, and every distinct galaxy testifies to a single truth – **the Creator's genius is inexhaustible, and His Majesty beyond measure.**

Gary Ezzo
Author

PREFACE

"Data isn't information; information isn't knowledge; knowledge isn't wisdom."

Ian Lowe[1]

"The facts are known but they are insulated and unconnected... The pearls are there but they will not hang together until some one provides the string."

William Whewell (1794-1866)[2]

The Information Age is transforming how we perceive life, the universe, science, and the very nature of information itself. Central to this shift is the advent of powerful tools, with the computer being foremost among them. A computer consists of a tangible physical platform, the hardware, which exists within a material realm that physicists call *Space-time*. In contrast, the software operates within this hardware in an intangible, non-material domain often referred to as *Cyberspace*, a term used for over three decades.[3] While software, through its code, delivers information via the hardware, the hardware alone cannot convey information.

Even so, the electronic circuitry that forms the material hardware is an essential *vehicle* within space-time, which *carries* the software that conveys the code, while the information itself is non-material. This is similar to the information code embedded in the DNA that specifies and operates every living cell. Equally, it parallels the entity that underpins the natural laws that underlie and sustain the Universe.[4] This non-material code is organised data that derives from quantum information. If this sounds strange, think of the chemical structures that construct the entire Universe, then take a step back to consider the entities that underlie the atoms within these chemicals, the protons, neutrons and electrons that form the atoms, then take a further step back to contemplate the constituents that underpin the sub-atomic particles, which at ultra-microscopic (Planck) scales emerge from within the quantum realm. Underneath all this, it is proposed there is information lurking that configures the lot![5]

But thinking informationally requires a reappraisal of our traditional understanding of information. According to the orthodox view, mathematical relationships describe the basis for our material existence. The physical world is an expression of these mathematical relationships, whereas information is a secondary, or derived, concept that characterises specific states of matter (such as an electron spin being up or down). Thus:

The orthodox view:
Mathematics → Physics → Information

However, physicist, Paul Davies points out that the laws of physics are simply information statements: they tell us something about the way the physical world operates. In this alternate view, information, manipulated according to mathematical principles, describes the laws of physics through which matter is derived. Thus, embedded information is the fundamental entity. Hence:

The information view:
Information → Laws of physics → Matter[6]

From this information perspective, the Universe can be viewed as an amalgam of material and non-material information-based entities which behave in one of four different ways, suggesting that there are four different forms or domains of information. Each will be examined separately in Chapter 7; but how did we reach this conclusion?

The formal study of information as a discrete discipline dates from the latter 20th Century but since most of life and especially the sciences depend to some degree on collecting, processing and/or applying information, the pragmatic utilisation of information dates from the beginning of recorded history. Thousands of Babylonian clay tablets attest to this in such mundane matters as keeping trade accounts. The active utilisation of information is a defining characteristic of the human mind.

Since Babylonian times, this has grown into an enormous field which is often perplexingly counter-intuitive. Some examples include:

1. Space-time is a single *four*-dimensional concept.
2. The Universe started from nothing (that we can detect) and yet is rich and vibrant. How?

3. The law of entropy decrees that everything is winding down, except life which (it is proposed) began as a single cell and does the opposite, becoming more complex!
4. Living things are constructed from non-living parts; the difference is profound but detecting a material difference has eluded us.
5. The human brain and computers have much in common except that the brain can be conscious while computers have never been so, but the mechanism that makes the difference has not been identified.
6. Einstein's Relativity Theory and Quantum Mechanics form the foundations of modern physics. Both have been proven to be correct and yet they are mathematically incompatible with each other. How does this work?
7. The Standard Model of Physics describes our universe as the result of random processes, yet it is underwritten by some of the most elegant mathematics known to humankind; further, scientists found the positron only after the mathematics directed scientists to look for it. How could a 'randomly' formed universe have 'known' this?
8. And perhaps most shockingly of all, the apparently 'concrete' material world seems to arise from the quantum domain that is blurry, wavy and smeared.

After 400 years of studying the Universe, scientists are still scratching their heads trying to understand and explain these and many other paradoxes.[7] Material science is yet to provide coherent explanations to these fundamental observations. Yet, this is not for the want of trying. The *annual* output of scientific research in the 21st Century is likely to total more than the previous three centuries put together and yet the last discovery that significantly extended our fundamental understanding of the Universe came with quantum mechanics a century ago. So, could we be looking in the wrong place? In this book it is proposed that instead of the Universe being fundamentally a material or solely a quantum entity, it is instead constructed from information that has resulted in the vital, astonishing universe that we see around us.

Materialism has done an excellent job at describing what can be found in space-time, which has led to exploration and discoveries of what lies beneath it in the quantum world but in this investigation into information we will be journeying even deeper into reality, into what underlies the quantum domain and has the potential to unify these previously counter-intuitive parts of our cosmos!

Indeed, as we pry deeper into the fabric of our universe, the more information we find. Walker, Davies & Ellis write: "*... information is often treated as a primary entity, as a 'thing in its own right' with a measure of autonomy; indeed, it is bought and sold as a commodity alongside gas and steel.*"[8] Responding to this by changing our frame of reference from being exclusively materialistic to being information-based gives us a new perspective on our universe which I have termed *Information Mechanics*.

Good science is based on sound evidence that has been tested. It uses robust logic to search for outcomes ('the fruit') that are beneficial in the real world. In doing so it provides *explanatory power* that guides what we do next.[9] If we are brave enough, we will follow the findings of this process. Unfortunately, as human beings we can insert what we would like to see rather than follow the evidence, which often ends up taking us down a rabbit burrow and getting lost. This is what modern science has done with materialism, by making it the be-all-and-end-all of science, about which Alistair McGrath wrote that we have been "*impoverished by* [science's] *reductionist agenda*."[10] However, by the end of this book we will have shone the new more powerful light of Information Mechanics on many of the more difficult-to-understand areas of science, which does provide greater explanatory power. So, let's begin!

Robert Wiles
October 2025

CHAPTER ONE

INFORMATION MECHANICS:

WHY WE NEED IT

"...as far as we can tell, the ultimate nature of reality is informational..."

– Luciano Floridi[1]

Since ancient times, human beings have gazed about the Earth when it is bathed in sunlight and endeavoured to understand this planet on which we are fortunate enough to live. During the day, the biosphere reveals an amazing array of colour, beauty and vitality that is everywhere around us but then as the sun sets, we can gaze up at the night sky where the darkness provides a lens to view the Universe, that we might wonder at the vastness far beyond this Earth and its cosmic origins.

However, our understanding of our cosmic origins has changed over the millennia, and the pace of that change has never been as great as over the past century. In 2013, it was estimated that human knowledge was doubling every 13 months[2] and accelerating! In the 21st Century, information has become one of the most sought-after commodities and criminal gangs, even hostile governments, will stop at nothing to steal it. As a result, our knowledge regarding all aspects of information is increasing; both for what it conveys as a valuable resource (the message) but also to optimise the entity that is the carrier of the information (the vehicle) to maximise its utility, and what we are learning has astounding implications!

1.1 – INFORMATION THROUGHOUT HISTORY

The first records of a scientific nature that seek to explain the structure of our universe have been traced to Ancient Greece. In the 5th Century BC, Leucippus (c.480-c.420 BC) along with his student Democritus (c.460-c.370 BC) devised a theory that envisioned our world as composed of indivisible atoms. This extraordinary insight eventually led to modern particle physics and Professor John Lennox observes: "*...laid a foundation for a materialistic view of the universe.*"[3]

Fast forward to the scientific revolution of the 17th Century, with the discovery of Isaac Newton's 'clockwork universe', when a deterministic, materialist way of viewing the world appeared to be locked in by the mathematics; so, by 1895 it was thought that 'the science was settled' on the mechanistic universe. However, at the dawn of the 20th Century came the shocking discoveries of quantum physics which continue to confound us today. The atomic particles that were thought to be tiny hard balls were discovered to behave like intangible waves rather than solid matter, depending on how we interrogate the evidence (data). The revered Newtonian *Classical Mechanics* with its comforting determinism had to make way for 'messy' *Quantum Mechanics* with its uncertain probabilistic characteristics. However, as we race through the 21st Century, a century on from Planck (1858-1947), Einstein (1879-1955) and Schrödinger (1887-1961), we observe that progress has stalled; the classical and the quantum still appear to be mutually contradictory and hence incompatible, despite the experimental data and the mathematics demonstrating that *both are correct*, while simultaneously appearing to prove each other wrong! This conundrum has troubled a substantial number of scientists, not least of all, Einstein, who devoted a significant part of his latter career to trying to resolve the mystery. Einstein's Relativity Theory extended our understanding of the materialism of classical mechanics but failed to bridge the chasm, across which lay quantum mechanics.

However today, a growing number of physicists are in no doubt that the foundation of the real world is quantum, not materialism.[4] Classical mechanistic physics is a useful shortcut in biology and medicine, meaning that we can usually avoid having to delve into the mystical, murky depths of the quantum world and most of us never have to think about it. NASA successfully sent Apollo 11 to the Moon and back in 1969 using Newtonian mechanics because its approximations to the quantum foundation made an insignificant difference, but reality still ultimately lies in quantum mechanics. So, materialism, as experienced in space-time, appears to be an emergent property of the weird quantum world which underpins it. This dichotomy has bamboozled even the best scientists for three generations but most fortunately for us, the scientific discoveries did not stop with quantum.

The past fifty years have witnessed an explosion of information, which has been paralleled by the growth of technology, the likes of which has never been seen before in human history. This has not happened in a vacuum. There is a long list of original thinkers who have contributed to the accelerating pace with which the information revolution has poured forth. Key scholars have included: Charles Babbage (1791-1871), Ada Lovelace (1815-1852), Norbert Wiener (1894-1964), Alonzo Church (1903-1995), John von Neumann (1903-1957), Alan Turing (1912-1954) and Claude Shannon (1916-2001) who have all made notable contributions.[5]

The term: *Information Mechanics* was introduced in 1974 by Frederick Kantor (1942–2020), an American physicist,[6] who was building on the ideas of Konrad Zuse (1910–1995), a German engineer and early computer scientist. In 1969, Zuse had proposed the concept of a computation-based universe in his book *Rechnender Raum (Calculating Space)*.[7] Following Kantor, this concept was taken up by Edward Fredkin (1934-2023) of MIT who preferred the term: *Digital physics*, which he coined in 1978, but later changed to *Digital philosophy*. Digital physics proposed that the Universe can be conceived of as a vast, digital computation device, or as the output of a deterministic computer program.[8] In 1989, physicist John Archibald Wheeler (1911-2008) elevated the visibility of the debate with the phrase: "*It from Bit*" at the Third International Symposium on the Foundations of Quantum Mechanics held in Tokyo: "*...in a broader sense, of the theme that every physical entity, every it, derives from bits* [of information]."[9] More recently, Lucino Floridi, Professor of Philosophy and Ethics of Information at the University of Oxford introduced the term: *Philosophy of Information*, suggesting that the topic now deserves the status of a discipline in its own right.[10]

Although physicist and electronics engineer Harry Nyquist (1889-1976) first referred to what was transmitted by telegraphers as 'information' in 1924,[11] it was not until after World War II that *Information Technology* began its meteoric rise. In 1948, Claude Shannon (1916-2001) and Warren Weaver (1894-1978)[12] described the *transmission* of information arithmetically. *Shannon information* describes mathematically, the process by which information can be moved or distributed and can be thought of as the "handling" or "processing" of information. Formally it is called *syntactic* information, which is a measure of the *quantity* sent. Think of it as an electronic postal service, describing how many packets of information can be sent down a telephone line. However, Davies & Gregersen comment: "*When the foundation of information theory was laid down by Shannon, he purposely left out of the account any reference to what the information means and dwelt solely on the transmission aspects.*"[13] Weaver subsequently referred to Shannon Theory as The Mathematical Theory of Communication (MTC) which: "*...deals with the carriers of information, symbols and signals, not with information itself.*"[14] So, with Shannon information we can know how many packets are being sent yet know nothing about what the packets contain and hence their meaning.

1.2 – RECOGNIZING THE SIGNIFICANCE OF INFORMATION

The latter 20th Century witnessed the ascendance and wide availability of computers, but this has not solved the meaning problem because, as Paul Davies noted: "*...the laws of physics merely shuffle information around ...*",[15] which means that although these natural laws can move information, they cannot create it nor understand it. The problem is that as computers operate in material space-time, they are compelled to abide by its deterministic laws that appear to negate any option to choose.[16] This inability of computers to create nor interpret the meaning of the information being transferred led some investigators to consider fresh ways as to how this problem might be overcome. Jacob Bekenstein (1947-2015), Professor of Theoretical Physics, wrote in Scientific American in 2003: "*Ask anybody what the physical world is made of, and you are likely to be told 'matter and energy.' Yet if we have learned anything from engineering, biology, and physics, information is just as crucial an ingredient... Indeed, [some physicists] regard the physical world as made of information, with energy and matter as incidentals.*"[17] Then, in 2006, Professor of Mechanical Engineering, Seth Lloyd went further, saying:

"...everything in the universe is made of bits. Not chunks of stuff, but chucks of information - ones and zeros."[18]

Discussing the significance of information, in 2011, James Gleick stated: *"Increasingly, the physicists and the information theorists are one and the same. The bit is the fundamental particle of a different sort: not just tiny but abstract - a binary digit ... more fundamental than matter itself."*[19] This was first enunciated by John Archibald Wheeler who wrote in 1989: *"**It from Bit.** Otherwise put, every **it** — every particle, every field of force, even the spacetime continuum itself — derives its function, its meaning, its very existence entirely — even if in some contexts indirectly — from the apparatus-elicited answers to yes or no questions, binary choices, **bits**.*

It from Bit symbolizes the idea that every item of the physical world has at bottom ... an immaterial source and explanation; that what we call reality arises in the last analysis from the posing of yes-no questions and the registering of equipment-evoked responses; in short, that all things physical are information-theoretic in origin ..."[20] So, in summary, the number, stature and clarity of the voices affirming that information underlies the Universe, is getting too loud to ignore.

1.3 – UNDERSTANDING THE CURRENT MATERIALIST PARADIGM

The problem is that, as Gleick suggests above, information is an abstract or non-material entity, so it cannot exist in space-time unless it is embodied, meaning that it is within or part of something else that is material. In 1964, Marshall McLuhan (1911-1980) famously coined the phrase: "*The medium is the message*", which may have been true in advertising at that time, but with information, the medium is distinctly different from and subservient to the message. The medium carries or transmits the message (the information) but the same information can be conveyed by many different media, such as in print, digitally by email, through the spoken word or via radio waves to name only a few.

The problem is that knowing how many packets of information are travelling around our material universe does not provide any meaning or significance to the *contents* of the packets; nor does it adequately describe or explain all that we observe. There are still many observations that remain unexplained within our materialist paradigm, such as:

1. That the Universe is ordered, logical and understandable. [Einstein 1935[21] – observe that chance usually leads to *disorder* or *turmoil* due to the Law of Entropy.]
2. That mathematics can describe so much of the Universe. [Wigner – *The Unreasonable Effectiveness of Mathematics*, 1960][22]
3. That much of the mathematics is elegant and relatively simple. [$E=mc^2$ Einstein 1905][23]
4. That life exists. [de Duve 2002][24]
5. That cellular machines are so efficient and so exquisitely constructed. [Alberts 1988][25]
6. That consciousness exists, and human beings are self-aware. [Chalmers 1996][26]
7. That fine tuning is so commonly found. [Lord Rees – *Just Six Numbers* 1999][27]
8. That the Universe can be described by information. [Floridi 2011][28]

Each one of these is extraordinary and deserves an answer. Answers are coming but the solutions require some radical rethinking regarding the construction and functioning of the Cosmos.

1.4 – THE CHANGING ZEITGEIST

There is a common thread of information and meaning that can be traced through these eight observations that cannot be explained by either classical or quantum mechanics. As a result, the zeitgeist or worldview (also called a *paradigm*) has been shifting. A worldview is our mental map of the Universe, as we *perceive* it to be. It can be thought of as the pair of glasses through which we see the World, though our worldview largely goes unrecognised and unquestioned, which is to say that we fail to notice the 'glasses' we are wearing when we view our world. Most significantly, the more closely our worldview aligns with the reality we experience, the better we can understand and negotiate life.[29] It is so fundamental that when it changes, it can be very unsettling, so we do not change worldviews very often. One of the biggest problems with our present worldview is that our currently accepted scientific theories cannot adequately account for *meaningful* information (formally called *semantic* information), which is ubiquitous in life, as found in the DNA code.

So, from the foregoing discussion, it would appear that the materialistic world that we experience in space-time is just *one part* of a more complex universe that emerges out of quantum mechanics, which itself (as we shall see) appears to derive from underlying information. Studying matter revealed material space-time but also uncovered the quantum world. However, even a century of research delving into quantum mechanics has not been able to knit these fields together so that we are able to understand the list of unexpected observations above. This suggests that we need another approach, one that is more fundamental than either classical or quantum mechanics and information is shaping up as a possible candidate. However, to do this we need to view the Cosmos through multi-focal lenses to bring space-time, quantum mechanics and information all into focus at the same time. As we shall see, this will provide greater explanatory power. This new focus can be understood better from within a new paradigm of *Information Mechanics*.[30]

1.5 – INFORMATION MECHANICS

So, what is information mechanics and how might it help? Information Mechanics involves the study of the various information systems that operate within the Universe, their generation, characteristics and applications which will be discussed over the succeeding chapters. Much research has been invested in Shannon information over the past seventy years due to its direct applicability to telecommunications, while semantic or meaningful information has been left behind as it has defied mathematical description and has lacked a direct commercial application. However, the development of personal computing from the late 1970s and the sequencing of the human genome in 2000 have brought the study of semantic (meaningful) information to the fore. Developments to-date have been slow though, because it requires a radical rethink in our understanding of the fundamental construction of the Universe, which Thomas Kuhn (1922-1996) called a *paradigm shift*.[31] History shows that these radical changes in the widespread understanding of science occur rarely, perhaps once in a century and usually take one to two generations before they are fully accepted, allowing time for the changes to ripple through all the branches of science as new students are taught the new ideas on entering the academe. The change is completed when this new guard rises to the top of their disciplines.

Information Mechanics positions information, rather than Democritus' indivisible atoms made of matter, as the foundational entity of the Universe. Today, viewing the world as being exclusively constructed from matter and energy is referred to as Materialism. To understand the differences between these two paradigms we use the tools of science. Science is the study of the *processes* by which the Universe is configured and operates. Unless we understand the Universe the way it really is, science will be hindered and unfortunately science is currently stuck with the incompatibility of classical and quantum mechanics. Although confidence in the Standard Model of Cosmology is being questioned, there is little consensus on where to go next. String Theory[32], M-Theory[33] and Loop Quantum Gravity[34] have been proposed but many physicists lack confidence in any of these solutions at this stage.

Another problem is that the prevailing doctrine of materialism cannot allow any exceptions if it is to survive, so there are arguments to 'demonstrate' that information is 'material'; however, even a superficial appraisal of the information code in DNA reveals that the code itself is immaterial. To exist in space-time, it requires a vehicle to carry it but as Tibor Molnar, Philosopher of Science at the University of Sydney states, information, along with mathematics are constructs or models that exist within our minds.[35] So, material objects, such as the complex molecules that make up DNA or the wooden blocks that the Duke of Devonshire has mounted on the wall at Chatsworth, his home in England, to record a fragment of his own genetic sequence can store the code in a material representation but the code itself is an entirely different thing.[36]

1.6 – INTRODUCING MIND

We previously noted that computers do not create information but can only shuffle it about (using pre-programmed algorithms). So, computers are an information management tool rather than generators of meaningful new information. Many scientists are currently working on *artificial intelligence* (AI) with the aim that AI can create original thought but at the time of writing, this has not been achieved. To formulate original ideas would appear to require mind or what Floridi terms: *'natural intelligence'* (NI).[37]

The human mind, manipulating meaningful information via consciousness, can conceive of radically different concepts to computers.

Materialist scientists maintain that mind is an *emergent* property of the active brain[38] which would suggest that our mental faculties would age in parallel with our physical brain substance but those of us who are old enough to have lived through five decades or more know that our mind does not consider that it is the same age as the body in which it resides, which can be confirmed by one brief glimpse in the mirror. Once we have reached adulthood, if we are healthy, our mind appears *not* to age in the same way as our physical bodies. Our powers of recall, speed of recognition and creativity tend to slow down but this seems to be due to the aging of our material brain substance or fabric, the physical 'hardware', rather than the mind which seems to operate more like a software application built from information.[39] Thus, mind appears to be a non-material entity running on a material platform. This is an example of materialism and Information Mechanics revealing their differences; yet these markedly different material and non-material components still work seamlessly together. Our intuition may suggest to us that a material explanation is the logical way to interpret what we experience in the World through our senses, but our senses have limitations and can deceive us; we may have an incomplete understanding, such as we might experience when we 'see' the mirage of an oasis in the desert. So, what might a complete understanding look like? The problem is that we are not sure yet but there are some clues that we will discover as we follow the information trail and dig deeper in Chapter 2.

CHAPTER TWO

EXPLORING INFORMATION

"...energy makes physical systems do things. Information tells them what to do."

– Seth Lloyd[1]

In Chapter 1 we discovered that information is a real entity that is integral to our Universe and that by following the trail of information we can discern that there seems to be more to the construction of our cosmos than just a material framework. In particular, the quantum components have often been overlooked as they appear to be incompatible with our familiar physical world from which most of our analysis begins. This suggests that there must be something missing, an underlying explanation that is able to link these two realms together. Having discussed some introductory issues around information and space-time, we now begin digging deeper.

2.1 – RE-ENVISIONING THE UNIVERSE

Quantum physicist, Carlo Rovelli, tells us that space and time are not fundamental but that they emerge.[2] So, the question is: From what and from where do they emerge? Presumably from something or somewhere else. Einstein told us that space and time are a single entity in our physical universe, so in Einsteinian mechanics (also called *Relativity Theory*) we need to treat them as one unit that physicists call *Space-time*, which exists in four dimensions: three dimensions of space and one dimension of time. Physics tells us that when we drill down to the ultra-miniature (much less than even the size of a quark), called *Planck scale*, below Planck length of 10^{-35} metres, matter loses its physicality and quantum mechanics then determines the rules. So, space-time would seem to emerge from the extremely tiny quantum world which is invisible to our senses. However, Tibor Molnar tells us that quantum mechanics is emergent too, which means that it is not fundamental either; then Molnar adds: "*Information is at the bottom of everything we know, but not at the bottom of everything.*"[2] The problem is that at present, we do not have a robust model to know what that would look like; information outside its carriers has no substance. At this ultra-microscopic scale, entities lose discreteness so that they are not so much a thing as a process, which Rovelli considers is a more helpful way to think about what is happening at the foundations of physics. So, he encourages us to consider electrons as events or processes rather than objects.[2] Now this is a very long way from my intuition that I am surrounded by discrete, physical entities, such as my desk and chair where I am writing and so it all requires a radical rethink! Perhaps a way to assist in our reappraisal is to look at the things that physics considers fundamental which include *space-time, matter, energy* and in the light of the above discussion, *information*.[3]

2.2 – ENLISTING THE TOOLS OF SCIENCE

As mentioned above, we will be using scientific principles and logic to discern the road ahead but first, we need to digress to explore how we can even conceptualise what to many of us is a visit to an alien world of non-materiality. In doing this, even finding the words to express what we visualise in our mind can be difficult. To help with this we construct models which Molnar describes as: "*...virtual concepts in our minds.*"[2] In doing this, Floridi reminds us that it is important to remember that the model is the medium, not the message.[4] Information is the content that is delivered by the model, which is its vehicle. Theories, algorithms and formulae are also models that vary in their completeness or ability to convey the Universe as it really is (that is, in actuality or truth). As such, they assist us in processing and understanding ideas but note that while a model can help us to 'see' an idea in our mind, if a theory or model is overrated or incorrect, instead of helping, it can convey the wrong idea or resist change when our knowledge has moved on, which can impede the advancement of science.

Stephen Hawking (1942-2018) & Leonard Mlodenow pointed out that we don't know (and may never know) exactly how the Universe works and certainly we cannot build a working replica on a scale to directly observe what is happening, so we build a virtual working framework of how we think it might be and how it might function, which we call a model. We then test the model by comparing it to what we can measure or observe in the real world and make predictions from the model that we would expect to see if the model is accurate.[5] David Krakauer, Professor of Complex Systems at the Santa Fe Institute explains that a good model must be:

1. explanatory,
2. predictive and
3. comprehensible to be useful.

The more accurately it does these things, the more powerful the model becomes.[6] However, unlike the model of a house we might like to build, a realistic model of the Universe requires both material and non-material (information) components, so it must exist in our minds as a virtual reality.[7] This lack of ability to *physically* test all the aspects of our model, requires us to step from science into philosophy which is no less rigorous but must operate under different rules due to the boundaries imposed upon us by our limitations in space-time. Sharon Dirckx of Oxford points out that when an idea cannot be tested scientifically there are three tests to assess its validity:

1. Is it internally coherent?
2. Does it have explanatory power, which is to say, is it externally relevant or useful?
3. Can it be lived? Is it feasible or just a nice idea?[8]

One of the purposes of this book is to address these questions with respect to information's role in the Universe so please read on! For readers who feel uncomfortable with the non-material components, see the discussion in the section on *mind* below.

2.3 – THE FUNDAMENTAL ENTITIES IN THE UNIVERSE

Newton's classical mechanics deals with masses (matter) and forces (energy applications) that make matter move in space-time.

Subsequently, Einstein, using his famous formula $E=mc^2$, showed that mass and energy are different forms of the same thing, which was proven when the first atom bomb was detonated. So, physicists now speak of *mass-energy* as being equivalent or one entity, in the same way that we speak of *space-time* as one concept. The Standard Model of Physics describes the sudden appearance of energy at the beginning of the Universe as it rapidly expanded. When I was at school, I was taught that mass and energy were located within the 'void of space' *inside* our cosmos, however, rather than our universe just being described as matter and energy within a vastness that is mostly vacant space, it is also described as lying within a *Space-time fabric*, implying that there is some extra dimensionality. We cannot see or feel this 'fabric' so its character is unknown but out of the seemingly empty void of space, *phantom particles* can appear and disappear. The space-time fabric appears to be the framework from within which space-time emerges and maybe through which the laws of the universe function, although there is no agreed understanding yet to explain how. Physicists have also discovered that much of the mass and energy that should be visible/measurable within the Universe is 'missing', so they have dubbed the missing elements, *dark matter* and *dark energy*. Could they also be squirreled away within the space-time fabric along with the phantom particles?

2.4 – PHANTOM PARTICLES

So, if phantom particles in space-time 'emerge', then they presumably originate from somewhere else, either outside or 'beneath' or 'within' space-time. This in not yet well understood but perhaps these phantom particles materialise then de-materialise again into what we might call the 'quantum domain'. Stephen Hawking and James Hartle developed a mathematical model to describe the beginning of the Universe and theorised that it came into existence through energy fluctuations in a quantum vacuum.[9] In school, we are taught that a vacuum is completely empty but Tibor Molnar clarifies this by pointing out that a vacuum contains *"nothing that you know"* and John Lennox confirms that a quantum vacuum is certainly not "nothing".[10] Professor William Lane Craig points out that subatomic particles such as virtual particles do not come out of nothing. The quantum vacuum is: *"a sea of fluctuating energy, an arena of violent activity that has a rich physical structure and can be described by physical laws."*[11] If this is where the phantom particles usually reside, then could the space-time fabric also be emerging from the quantum domain? This is yet to be determined.

However, this begs the next question: from whence came the quantum vacuum? You will remember that Molnar told us that *"information is at the bottom of everything we know"* so it seems possible that the quantum domain itself is specified by information beneath it. Paul Davies writes that this idea was first expressed by the eminent American physicist John Archibald Wheeler in 1989 when he stated: *"Tomorrow, we will have learned to understand all of physics in the language of information."* Then in 2001, *New Scientist* printed an article by Professor of Physics, Hans Christian von Beyer, entitled: *'In the beginning was the bit'* where he recounts the work of quantum physicist and Nobel Laureate, Professor Anton Zeilinger of Vienna, who attributes quantisation, among other quantum effects, to the spin on an electron being either up or down and hence carrying one bit of information. This suggests that information is more fundamental than mass/energy.[12] Widespread acceptance of these ideas by the broader scientific community has been slow but after nearly a century of research into quantum mechanics without resolution of the classical-quantum conundrum, information offers a logical avenue for further enquiry.

So, returning to Molnar's comment about information being at the bottom of everything we know, it is interesting to note that he finished the sentence by adding *"but not at the bottom of everything."* By this he appears to be implying that there must be something else to generate the information that formulates and/or sustains the quantum domain. In a later chapter we will explore what is required to generate information. However, before we do that, we need to discuss information itself.

2.5 – WHAT IS INFORMATION?

If information is the most fundamental entity that we know of in the Universe, then what is it? This is not an easy question to answer. Floridi points out that: *"Information is still an elusive concept."*[13] Even knowing how to define information is a challenge. Molnar points out that information means different things to different people. In 1969, British physicist, Donald MacKay (1922-1987) wrote that: *"information is a distinction that makes a difference."*[14] The *Cambridge Dictionary of Philosophy* defines information as: *"an objective (mind independent) entity. It can be generated or carried by messages (words, sentences) or by other products of cognizers (interpreters). Information can be encoded and transmitted, but the information would exist independently of its encoding or transmission."* (Audi, 1999).[15] This definition suggests that information is manipulated data[16] that conveys something; however, it implies a narrow interpretation that ignores the entity that underlies the physical elements of space-time (although this too may ultimately prove to be consistent with the Cambridge Dictionary of Philosophy definition, depending on our definition of 'cognizers').

A broader classification could describe information as constituting *structures*, such as physical entities 'that are' and information containing *systems* 'that do'. Much of what we have traditionally thought of as information falls into the latter category, such as semantic information that ascribes meaning or a set of instructions to build an entity, such as the DNA code. However, as we have previously seen, it can also be the *mathematical* (algorithmic) packaging-type syntactic information described by Shannon Theory that measures or distributes. Mathematics is a structured set of rules; think of it as a formalized operations system or language that defines how information may be legitimately manipulated. Mathematics is therefore an information meta-system that operates outside or above other information and indeed it is very useful for Shannon information. Both information and mathematics involve non-material entities which we manipulate in our mind to create a virtual reality within our consciousness and that (at a different level) probably also facilitates the functioning of our mind itself, which is discussed in Chapter 6. The functions of information are illustrated in Figure 2.1.

FIGURE 2.1: FUNCTIONS OF INFORMATION

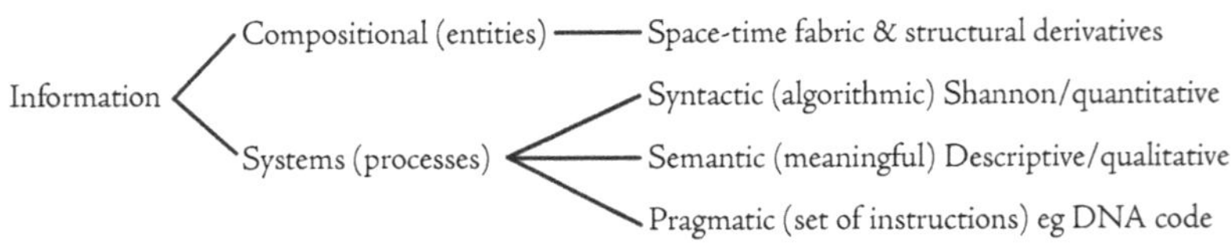

So, in summary, information defines a relationship that creates order and/or structure that has the potential to produce a functional or meaningful outcome. Order is an arrangement of elements that impart an objective or entity. The disruption of these informational relationships leads to disorder, which physicists term *entropy*. Most commonly, information is recognised as a series of instructions that direct a process, such as a recipe or mathematical algorithm. So, the organisation of information is the opposite process to the natural increase of entropy that is observed in the Universe. As such, entropy can be seen as a measure of the degree to which the components of the Universe are ordered or structured. However, being a non-material entity, in space-time, it must be stored in another entity that may be material, such as the chemicals that form Deoxyribose Nucleic Acid (DNA), or it can be a set of ideas in your mind such as the virtual reality models outlined by Molnar above. The problem though, is how our minds can 'see' these concepts to process the ideas, which requires that we are conscious, to store this information and then retrieve it again. This is yet to be fully understood, however there is a lot of work that is currently being undertaken, so now is a good time to take a closer look at the mind.

2.6 – THE SIGNIFICANCE OF MIND

Even if our research into mind cannot understand the 'how', we can discern the 'what'. Some of the non-living material carriers of information, along with the ability of computers to manipulate information have been discussed above, so now it is time to explore the role of mind in managing information. Many aspects of mind are still very poorly understood, such as consciousness, self-awareness, memory, creativity and understanding. To appreciate meaningful information requires consciousness as a platform to perceive it, memory to provide context and understanding to assign relevance. So, to conceptualise meaningful information requires a conscious mind.[17] Contrast this with computers that can shuffle or process meaningful information but cannot create it unless intelligently pre-programmed to do so. Current AI can mix and match existing information that is sourced from elsewhere, such as the internet. However, Marks et al point out that this ability is context specific, such as playing chess but that same chess program will not also write a novel.[18] Some computers have been programmed to write novels but require appropriate algorithms to be specifically written for that task. 'Machine learning' is in the same category, where large computers with sufficient programming can source data from the internet (which is a vehicle for information storage and retrieval) to extend their information base to create a 'new' arrangement or narrative but no computers are devising new hypotheses and directing novel research to turn into papers for publication in scientific journals based on *original ideas*. (At least not yet.) Currently, artificial intelligence (AI) can only *mimic* the output of the human mind, which Floridi dubs *natural intelligence* (NI) but cannot emulate the phenomenon of consciousness. 'Eureka' moments come from conscious intelligent minds. Floridi explains this difference as: "*... AIs can identify and process only data (uninterpreted patterns of differences and invariances), whereas NIs can identify and process mainly informational contents (... well-formed patterns of meaningful data).*" He concludes: "*AIs are data systems whereas NIs are information systems ...*"[19]

Virtually everything that is living seems to have some 'sense' of its environment including bacteria and plants as they actively seek nutrients and light from their surroundings to provide resources and energy to survive and they have basic information systems which conduct their day-to-day activities (see Chapter 6). Conscious reasoning and dynamic memory, however, are found only in vertebrates that have a sufficiently large brain, including birds, mammals and primates. Learning and creativity are generally exhibited in proportion to brain size but understanding seems limited to large vertebrates and self-awareness to human beings alone.[20] Understanding requires the consideration of context which requires memory, so requires extra processing power. Memory seems to be a record of past informational events which can be recalled at will. The creation of intricate meaningful information such as novel computer code or prose however needs the sophistication of a human mind, and no other living species seem to record complex messages for others to comprehend. Some animals such as dogs leave their scent on trees to mark their territory but this is a pre-specified process or 'instinct' that seems to be 'hard-wired' rather than a creative informational process such as writing a poem on that same tree.

These observations have great significance when we contemplate the informational content of DNA or the laws of the universe and how their meaningful information sequences may have arisen. The origins of these entities have long been debated and will be discussed in detail in subsequent chapters. What is clear is that the inorganic materialistic processes that have been described to date have significant problems in accounting for the origin of meaningful information but by using the ideas within Information Mechanics, we will show that solutions can be deduced.

CHAPTER THREE

THE CHARACTERISTICS OF INFORMATION

"Information is information, not matter or energy. No materialism which does not admit this can survive at the present day."

– Norbert Weiner (1948)[1]

For many of us, the field of information is a novel discipline, so in exploring this new subject, there is some specific nomenclature that will help us to study and understand it. It was previously noted that information has different functions in its myriad settings. These various functions entail divergent characteristics. It can be *quantitative* - as in Shannon's MTC, counting the information packets transmitted or *qualitative* - the meaningful contents of those packets. Shannon perceived a mathematical similarity between information distribution and the degree of order/disorder in the Universe, which physicists call entropy,[2] so information content is often discussed in terms of entropy but Jacob Bekenstein noted: *"Shannon entropy does not enlighten us about the value of information, which is highly dependent on context."*[3] Shannon-type quantitative information is often called syntactic whereas meaningful qualitative information is termed semantic. Some of these differences are summarised in Figure 3.1.

FIGURE 3.1: CHARACTERISTICS OF INFORMATION:

Note that the characteristics of compositional and pragmatic information (see Figure 2.1) are semantic in nature.

INFORMATION

QUANTITATIVE	QUALITATIVE
Stochastic (statistical) = MTC	Semantic (thematic)
Conveys data	Conveys information
Lacking meaning / context	Meaningful (conveys a message)
Functional (eg packet transmission)	Purposeful (eg defines entities)
Mathematically manageable (Shannon)	Indescribable mathematically
Compressible	Incompressible
Can form naturally	Must be generated intelligently

3.1 – THE INFOSPHERE: HARDWARE, SOFTWARE & INFORMATION

R.Z. Sheppard outlined the ubiquitous yet almost invisible nature of information in 1971 when he wrote: "*In much the way that fish cannot conceptualise water or birds the air, man barely understands his infosphere.*"[4] The infosphere is the entirety of all information. Luciano Floridi elaborates: "*... we are not standalone entities, but rather interconnected informational organisms..., sharing with biological agents and engineered artefacts a global environment ultimately made of information, the infosphere. This is the informational environment constituted by all informational processes, services and entities...*"[5]

Within the infosphere, there are multiple levels of information. The easiest way to visualise this is to consider a digital computer. The first requirement is some form of hardware which is a material entity that is the platform on which the non-material software runs. Running on the hardware is a machine language which interfaces with it; an operating system sits on top of these and then an applications package, such as a word processing programme performs the task. After all this, an output interfaces with you, the user. Table 3.1 shows the equivalent operation levels within the human brain. From these examples, it can be seen that the information can be intrinsic to the entity, such as the specifications for the composition of silicon in the microchips or extrinsic such as an applications package on your phone or an idea conveyed to you by a friend.

TABLE 3.1: HARDWARE AND SOFTWARE: COMPUTERS VERSUS HUMAN BEINGS

Note that both the hardware and the software, whether for computers or living organisms, is ultimately information. However, for computers, the operator is external while for the mind it is internal.

	Computers	Human Brain
Hardware	Computer components	Nerve cells and supporting tissues
Software	Machine Language	Biochemistry
	Operating system	Consciousness
	Applications packages	Speech or thinking
Operator	Extrinsic (output to the user)	Intrinsic (internal to the mind)

3.2 – THE FUNCTION OF INFORMATION

Walker, Davies & Ellis suggest that: "*… the life sciences,…treat information as the currency of activity, the 'oil' that makes the 'biological wheels go round.'*"[6] However, this functionality does not seem to be limited to the life sciences and is found throughout our entire universe. Information constitutes the instructions which specify the 'levers' that channel energy as it makes matter move; think of chemical bonds or the natural Laws of the Universe. However, as information is a non-material entity, it requires a vehicle to be transferred through space-time, the material part of our universe; this might take many different forms, such as a book, an optic fibre cable or a sonic wave through the atmosphere. So, it can be seen that matter, energy and information are all necessary for the effective functioning of the Universe.

Now, although information is the basis of the *physical* functioning of the Universe, it has even more significance for entities that are a composite of both *material* and *non-material* parts that combine to provide or convey meaning, such as your smart phone or computer. The true value of a computer is not in the hardware, although that is indispensable, but it resides in the software which conveys the meaningful information that makes it 'come alive.' This is true of much of the Universe but is most powerfully seen in the *mind-brain complex* which is discussed in Chapter 6. So, Walker et al suggest that even though information is intangible, yet it is real, as it has causal power that has the capability of changing outcomes and hence make a difference in physical space-time.[6] Therefore we need to take a closer look at this meaningful information that literally changes our world.

3.3 – MEANINGFUL INFORMATION

Nancy Pearcey writes: "*...information theory tells us that the medium does not write the message.*"[7] As an illustration of this, think of your computer hardware which cannot write its own software, let alone devise its information output. Some computers are being programmed to write new software and generate novel information; however, it is the information in the existing software that achieves this result, not the hardware in isolation. Further, meaningful output requires an intelligent source such as the human mind to generate the original software. From this we can see that meaning derives from the non-material; ultimately this means mind – see Section 10.3.

Similarly, random data alone cannot accomplish objectives either. Data is a resource that is not yet meaningful and yet it constrains meaning. It says, this, not that. This raw input is of minimal value until it is transformed into information by the provision of further information that delivers context, although often yet further processing is required to create knowledge that delivers structured meaning, which can then be used to achieve a desired outcome. Floridi points out that this occurs via relationships which define the way data works together so, meaningful data is relational.[8] John Lennox notes that as a result, "*...the meaning of a text is dependent on its context.*"[9] An isolated piece of information has no intrinsic meaning in and of itself which Floridi identifies as the *Data Grounding Problem.*[10] So, meaning is an emergent property of information where the interplay of differing facts or pieces of information (data) generate significance. As a result, information can only be verified by consulting further information. The first piece of information provides context that makes the following finding noteworthy. As information is complexed in larger and more sophisticated ways, the meaningful content increases. This applies to all areas of the Universe as can be seen in the examples in Table 3.2.

TABLE 3.2: THE CONSTRUCTION OF MEANINGFUL INFORMATION

	Language	Mathematics	Technology	Biology	Mind
Components	Letter	Number	Bit	Base	Event
Code	Syllable	Set	Nibble (4 bits)	Codon	Observation
Bundle	Word	Equation	Byte (8 bits)	DNA string	Thought
Thread	Phrase; sentence	Method	Algorithm	Gene	Idea
Fabric	Prose; poetry	Theorem	Programme	Tissue	Concept
Complex entity	Story	Treatise	System	Organism	Vision

Meaning is layered. Observe that at the *Component Level*, there is no significant meaning within the entities in the table, while the *Code Level* still endows negligible if any meaning, but at the level of *Information Bundling*, meaning starts to emerge. Although DNA is composed of a string of DNA bases, its power emerges when it is grouped into genes which then have purpose and hence can carry meaning that provides causal power. Genes can be thought of as 'groups of words' or 'phrases', written in the language of DNA. So, the *minimum* threshold for significant meaningful information content can be seen to be at the level of bundling, which is the 'word'. Each word is a little container of meaning. Lennox observes: "*...what has become more and more apparent ... [is] the central role played by 'word' in the shape of information at all levels in the sciences: physics, chemistry, bio-informatics, and computer science. ...we are increasingly learning, this is a word-based universe.*"[11] Braiding the bundled information together to create an information tapestry can rapidly convey rich, complex meaning such as stories, algorithms or organisms. Note that this is the way information is layered or *structured*. The way that information is generated will be addressed in Chapter 4.

3.4 – VERIFYING MEANING

Facts are deemed to exist after the veracity of a piece of information has been confirmed or tested over time. We say the facts represent the 'truth'. Floridi notes that in Testing Theory: "*...for a given model under test... there is a reference model...also known as an oracle, because it is assumed to be an infallible source of truth...*"[12] Identifying such oracles of 'truth' in the real world is not that easy. In testing theory, the oracle is *defined*, which is to say, it is specified by the scientist devising the test but in space-time, unless there is a relevant mathematical proof or the omniscient word of God can be applied, there is no infallible source of existential truth, so we have to use scientific principles to work it out.

In practice, *truth* emerges from the authentication of the accuracy of information over time. It is a measure of how closely our ideas or concepts in question align with the functioning of the Universe and hence the likelihood that we can use a piece of information to accurately predict the future course of events. Thus, truth refers to reality – what really is – which can be either physical in space-time or meta-physical in the quantum domain. *Knowledge* involves comprehending the value of a piece of information which permits its effective use in the future. *Understanding* requires the accurate interpretation of data or events through the application of context which involves the utilisation of prior knowledge to appreciate or perceive the significance or usefulness of a piece of information. This involves mentally apprehending the implications and/or relevance of the information in a way that permits us to effectively deal with the circumstance. *Wisdom* emerges from the sensible or most appropriate application of knowledge or understanding that has been found to be effective or fruitful over time, which permits good judgement.[13] It can be seen that the acquisition of knowledge, understanding and wisdom in the absence of a defined oracle requires an incremental stepwise process to discern the accuracy of the data in question. This is the engine room of philosophical reasoning and the scientific method and the reason why philosophy and science usually progress one minute step at a time. Each new piece of information needs to be triangulated off multiple other data points to confirm its validity. Note that each of these is an abstract term and hence non-material and requires a functioning mind.

3.5 – INFORMATION AND TIME

Information is independent of time. This is clearly demonstrated by the hieroglyphics on the pyramids in Egypt which still convey the same information encoded by their authors more than four millennia ago. The stone (vehicle) in which they are etched may be a little worse for wear, but the message is identical. This means that information of itself cannot tell you how old it is unless the author included a date stamp or other time dependant information. In practice, we date information from its physical carriers such as the parchment, inks or pigment used to record it. In the same way, the existence of information does not necessarily explain its causation unless the information also documents the author.

3.6 – THE IMPORTANCE OF THE QUANTUM WORLD

After Isaac Newton (1643-1727) published his landmark work on the laws of motion in 1687, the Universe was realised to not only be described by elegant and predictive mathematics but that the mathematics seemed to rigidly bolt the future behaviour of the Cosmos to its past in what became known as the 'Clockwork Universe'. So, once the Universe began, if you had a big enough computer with sufficient information about the state of all the atomic collisions, to calculate their outcomes, the entire future of the Universe could theoretically be accurately predicted ad infinitum. This predictability (which surprised Einstein)[14] results from stable natural laws that appear to be locked into the space-time fabric, creating certainty that permits logic but denies choice. This is called *determinism.*

This deterministic understanding of physics persisted until the early 20th Century when the discovery of quantum mechanics profoundly changed our understanding of the Universe and turned the world of physics upside down! Paul Davies and John Gribbin explain that when physicists drill down to the building blocks of our physical universe, our apparently firm physical reality is not what we intuitively infer from our space-time surroundings but that the *"...solid matter dissolves away, to be replaced by weird excitations and vibrations of invisible field energy. Quantum physics undermines materialism because it reveals that matter has far less 'substance' than we might believe."* This has resulted in the discarding of the deterministic Newtonian concept which has been *"replaced by a shadowy and paradoxical conjunction of waves and particles, governed by the laws of chance, rather than the rigid rules of causality."*[15] As discussed in Chapter 2, this phenomenon only supervenes at the ultra-microscopic scale which many of us find hard to contemplate. Most people, including most scientists, conveniently ignore the quantum world most of the time as it rarely, if ever, affects us directly. However, when we consider the mechanisms that underlie the functioning of the Universe it becomes vitally important. This will be discussed further in an investigation of the mind-brain complex in Chapter 6.

CHAPTER FOUR

SURPRIZING OUTCOMES

"There is no known law of physics able to create information from nothing."
– Paul Davies[1]

Over the past few chapters, we have encountered a number of features of information and their importance to our life here on Earth. Many of them are intuitive, some confusing and others are not what we were taught to expect, in school or even at university. In this chapter we start to unpack some of the surprising discoveries that emerge from the analysis of information.

4.1 – OUR MATHEMATICAL UNIVERSE

Since Pythagoras (c.570-c.495 BC) and Plato (c.427-348 BC), philosophers and scientists have wondered at the simplicity and predictability of the laws of the universe and the role that mathematics plays.[2] Two millennia later, Galileo (1564-1642) observed: "*The great book of nature can be read only by those who know the language in which it is written, and this language is mathematics.*"[3] This has been demonstrated by many, including Johannes Kepler (1571-1630) and Isaac Newton who uncovered the power of mathematics to make exceedingly accurate predictions about the movements of planets and the behaviour of our universe. Over the intervening centuries, we have been enthralled with the ability of mathematics to aid our understanding of how things work but although theories and formulae can reveal structured information, which provides an explanation that can extend our understanding, in and of themselves, these theories and formulae have no ability to create nor change physical space-time. They describe behaviour but cannot enact it.[4] For physical change there needs to be a mechanism such as the laws of the universe (like gravity), whose mechanisms and origins are still very poorly understood. These will be explored in the next chapter. However, this does not mean that the mathematics is not powerful. Fascinatingly, in 1931, Paul Dirac (1902-1984) predicted the existence of the positron (anti-matter) purely from the mathematics that describes the Universe. Oxford Professor of Physics, Frank Close commented: *"That to me is one of the great profound mysteries of science, ... It was a case of mathematics knowing about the Universe before we did."*[5]

4.2 – A QUANTIZED COSMOS

So, information appears to be emerging as the fundamental entity in the Universe and mathematics is the language that describes its behaviour. However, this is only the beginning of the story. Information can be digital (numerical) or analogue (such as text). In 1901, Max Planck (1858-1947) published a landmark paper on *quantisation* that shocked the scientific world, as the Universe had appeared to be continuous or analogue in character until then. What began with black body radiation in 1901, progressed to light and now to much of the Universe. James Gleick explains: *"Reality is quantized because information is quantized. 'The bit is the ultimate unsplittable particle.'"*[6] Clara Moskowitz, Associate Editor of *Scientific American* wrote in 2016: *"Lately scientists have begun to question... conventional thinking and speculate that space - and its extension according to general relativity, spacetime - is actually composed of tiny chunks of information. These chunks might interact to create spacetime and give rise to its properties such as the concept that curvature in spacetime causes gravity."*[7]

The astounding conclusion is that our universe is digital, which when we look around our planet seems counter intuitive. However, Rovelli writes that Nobel laureate, Paul Dirac viewed the world as not made of things but constituted of an abstract mathematical structure that shows us how things appear and how they behave when we encounter them[8] and in Chapter 5 we will also see that living cells are not only digital processors but much of their functioning is also digital. But first we must examine how the Universe came to be in a fit state to support life because most of our universe is very hostile to living things.

4.3 – THE UNIVERSE IS FINELY TUNED FOR LIFE

Over the past two centuries, many scientists have assumed that the Universe and the life we observe on Earth are the result of an enormous series of unplanned events. However, in the past seventy years, the discovery that so much of the Cosmos requires exquisitely precise specification just for you to be able to exist, to read these words, and that so much is also described by mathematics, is inducing increasing numbers of scientists to ask questions regarding the more than one hundred fortuitous co-incidences that make life possible, such as:

1. That our Universe seemingly appeared out of nothing with perfectly balanced laws;
2. That it expanded at exactly the right speed to form our stable, functional cosmos;
3. That our Solar System is structured to support life on the Earth;
4. That life developed bewildering complexity despite the Law of Entropy (unravelling);
5. That complex life developed consciousness to experience this world;
6. That human beings developed self-awareness to contemplate all the above!

Each event in this list requires new complex information to specify it but the new information must in addition be tightly specified. There are so many factors, such as the charge on an electron, the relationships between the four fundamental forces in the Universe, the properties of carbon and water that permit life, the size of the Earth to retain water and its distance from the Sun so that water is liquid. The wavelength of the Sun's light to permit photosynthesis, the tilt of the Earth for seasons and the relatively large size of the Moon to stabilize the Earth's rotation so that the oceans do not freeze over. In 1999, Astronomer Royal, Martin Rees published his: *"Just Six Numbers"* which describe the precision of six mathematical constants that define the form of the Universe that permits the human race to exist. Further, if any of these numbers was changed even slightly, then life as we know it *could not* exist.[9] Physicist, Paul Davies gives as an example, the ratio of the electromagnetic force-constant to the gravitational force-constant which cannot be altered by even one part in 10^{40} for there to be both large and small stars formed in the Universe, which are both essential for the current Standard Model of the Universe to work.[10]

In 2008, *New Scientist* published an article by Stuart Clarke entitled: "*Unknown Earth: Our Planet's Seven Biggest Mysteries.*" In this article Clarke asks the question: "*How come earth got all the good stuff?... We know that its distance from the Sun provides the right amount of heat and light to make the planet habitable... Without the unique mix of carbon, hydrogen, nitrogen, oxygen, phosphorus and sulphur that makes up living things, and without liquid water, life as we know it could not have evolved.*"[11]

4.4 – THE EVOLUTION OF SCIENCE

This very delicate balancing act was very powerfully demonstrated in the 1950s when Sir Fred Hoyle (1915-2001) discovered that carbon only exists in significant amounts in the Universe through a freak coincidence that requires the simultaneous collision of three helium nuclei that requires a phenomenon called resonance, an harmonic vibration that is not common but just happens to occur in this reaction. Without this occurrence, there would not be enough carbon for life. Paul Davies wrote: "*Hoyle himself was deeply impressed by this discovery. 'It looks like a put-up job,'... A similar conclusion was reached by the Princeton physicist Freeman Dyson:* [who stated] *'In some sense, the universe knew we were coming.*"[12] In fact, Hoyle went further stating that it looked as if: "*a super-intellect has monkeyed with the physics as well as with chemistry and biology*" and that "*there are no blind forces in nature worth talking about.*"[13]

Two quotes from Hoyle illustrate the change in thinking within the scientific community that occurred over the latter part of the 20th Century. In 1951, he wrote: "*I have often seen it stated that our situation on earth is providential. The argument goes like this. It is providential that the earth is of the right size and at the right distance from the sun. It is providential that the sun radiates the right kind of heat and light. It is providential that the right chemical substances occur on earth. A long list of this sort of statement could be compiled, and to some people it looks as if there is indeed something very strange and odd about our particular home in the universe. But I think that this outlook arises from a misunderstanding of the situation. Because if everything was not just right we should not be here. We should be somewhere else (or more likely, nowhere).*" However, thirty years later, in 1983, in light of all the intervening discoveries, Hoyle wrote: "*Such properties seem to run through the fabric of the natural world like a thread of happy coincidences. But there are so many odd coincidences essential to life that some explanation seems required to account for them.*"[14] In summing up all these findings in 1988, Paul Davies wrote: "*It seems as though someone has fine-tuned nature's numbers to make the universe... The impression of design is overwhelming.*"[15]

4.5 – IMMATERIALITY

In a famous paper published in *Physics Today* in 1991 entitled: '*Information is physical*', and a similar one five years later entitled: '*The Physical Nature of Information*', Rolf Landauer (1927-1999) wrote: "*Information is not a disembodied abstract entity; it is always tied to a physical representation. It is represented by engraving on a stone tablet, a spin, a charge, a hole in a punched card, a mark on paper, or some other equivalent.*"[16] However, each of these lies in space-time where the vehicle that carries the information is certainly physical but as John Lennox confirms: "*...the information is not only invisible, it is immaterial.*"[17] Information only *appears* to become physical when embodied within a space-time vehicle which by necessity must be a physical entity. John Lennox points out that: "*This misunderstanding has led to a great deal of unnecessary confusion that might have been avoided*"[18] and quotes the Oxford philosopher C.G. Timpson who describes the vehicle as: "*a token of a piece of information*" rather than the information itself.[19]

For the same reason, this phenomenon is also found in living things. Evolutionary biologist, George Williams (1926-2010) relates: "*The DNA molecule is the medium, it's not the message.*"[20] So, in the same way that information in inanimate physical entities does not make the information physical, a piece of genetic information stored in a Deoxyribose Nucleic Acid (DNA) molecule can alternatively be communicated by an engraving upon a stone tablet, holes in punched cards, marks on paper, or some other equivalent. Our genetic code that specifies who we are is therefore an independent entity to the DNA chemicals that are our genetic code's physical 'home' as we live and dwell in space-time. This distinction has been overlooked until relatively recently but is vitally important for us to properly understand our true 'makeup' or 'essence' as human beings.

Some materialists, such as David Armstrong (1926-2014)[21] and John Searle[22] have argued that thought and hence our mind are just a product of what our physical brains do but as will be seen in Chapter 6, consciousness, mind and thought seem to be the product of non-material processing delivered by software entities that run on the physical platform that is our brain, so once again, information in our thoughts appears to be a non-material entity that is hosted by a physical platform in space-time. Most significantly, as Nancy Pearcey previously pointed out, the physical medium never writes the code for the non-material message it carries. The conclusion that can be discerned from the above is that information is immaterial and distinct from its physical vehicles. Further, as the immaterial information that can be identified in the Universe (including our minds) cannot derive from its physical carriers, physical entities cannot explain the *origins* of the information nor mind itself either. To understand the origins of information we will have to look elsewhere.

4.6 – INFORMATION GENERATION

Quantum mechanical engineer, Seth Lloyd observed: "*Everywhere you look, you see immense variation and complexity. Why? How did the universe get this way? We know from astronomical observation that the initial state of the universe…was extremely flat, regular, and simple. Similarly, the laws of physics are simple:… Simple laws, simple initial state. So where did all of this complexity come from? The laws of physics are silent on this subject.*" Lloyd then answers his own questions by asserting that: "*… a computing universe necessarily gives rise to complexity … a universe that computes must, with high probability, give rise to a stream of ever-more-complex structures.*"[23]

In practice, however, we observe that meaningful information needs to be *actively* generated using a *pre-programmed* device. Bernd-Olaf Küppers, Professor of Natural Philosophy at the Friedrich Schiller University at Jena, Germany writes: "*…Information conveying meaning cannot be compressed without change in, or even loss of, its meaning… This is the case for any meaningful information …Viewed technically, this means that no algorithms – that is, computer programs – exist that can extrapolate arbitrarily chosen parts of the message and thus generate the rest of the message. But if there are no meaning-generating algorithms, then no information can arise de novo. Therefore, to understand a piece of information of a certain complexity, one always requires background information that is at least of the same complexity. This is the sought-after* *answer to the question of how much information is needed to understand some other information. Ultimately, it implies that there are no 'informational perpetual motion machines' that can generate meaningful information out of nothing.*"[24] John Lennox neatly expresses this as: "*Hence, no Turing machine can generate any information that does not either belong to its input or its own informational structure.*"[25] This concept can be summarized as the Golden or *Universal Rule of Information*, which states that meaningful information cannot appear at random or by chance; it needs to be actively generated. However, as original information cannot be generated by a computer (Turing machine) without prior programming, the minimum requirement for new meaningful information is an active mind. Further, as meaningful information must be actively generated, it comes at a cost, both to generate and to maintain or conserve.

4.7 – CONSERVATION OF INFORMATION

Once information has been generated, it can be conserved in space-time if a suitable storage medium can be found. However, the *No-hiding Theorem* states that even if a physical process seemingly destroys an original piece of information in space-time (in practice, its vehicle), the information is never completely lost due to the behaviour of quantum mechanics (see Chapter 8). This was proven by quantum physicists, Samuel Braunstein and Arun Pati in 2007[26] and successfully tested in 2011.[27] This is a mathematical proof that has implications for the black hole information paradox and in fact any process that loses information in space-time. Black holes will be discussed further in Chapter 8. However, this implies that quantum mechanics has an information storage function that has not yet been described.[28]

4.8 – INFORMATION AS THE FUNDAMENTAL ENTITY

The foregoing discoveries suggest that information has a unique place in our universe. Physicist, James Glattfelder writes that information is neither matter nor energy but pre-dates them both. "*The newly forming ontology [understanding regarding our existence] speaks of space-time being emergent and information [being] the only fundamental entity in reality.*"[29] Floridi also concludes: "*Although a very old concept, information has finally acquired the nature of a primary phenomenon...*"[30]

This can be tested by taking any topic or entity in the Universe and analysing it for its information content. Sometimes it is obvious and sometimes it is hidden but the information content can be discerned with some exploration. Take emergence as an example. Emergence is the phenomenon where something new and unexpected appears out of the functioning of the existing system, that would not initially be anticipated from the appearance of that entity. We have already seen the emergence of phantom particles from the void of space in Chapter 2 and space-time also seems to emerge from the quantum realm, while John Searle[22] implied that mind is an emergent property of the physical brain. Closer to home, Richard Dawkins described word-processing as an emergent property of computers[31] while 'wetness' (from surface tension due to hydrogen bonds) is the unexpected result of combining hydrogen with oxygen in the appropriate proportions to produce water molecules.[32] In each of these cases, the emergent property is the result of the expression of latent information that was previously hidden within the entity. Sometimes there is the subtle application of new information that initially goes unnoticed such as the hydrogen bonds in water. John Lennox writes: *"Thus the 'emergent' picture is only derived from the simple … by the cost of a hefty additional input of information in terms of programming effort and intelligently designed hardware. It does not come 'free.'"*[33] Therefore, emergence requires an information source that is able to generate that additional information. This same principle applies to the formation of our universe, along with the laws that govern it, so the Universe could not have generated itself either. Physicist and engineer, Professor Emeritus Edgar Andrews describes how before the Cosmos, there was nothing,[34] so there must have been an enormous injection of exquisitely crafted information for our current universe to have come into existence.[35]

So, in summary, a range of physicists, engineers, mathematicians, information scientists, biologists and philosophers tell us that information is now considered to be a primary, if not *the* primary entity in the Universe, however, the generation of that information seems to require a pre-existing mind or consciousness. In 2021, John Lennox concluded: *"I think the time has come to accept … the existence of an 'information first' universe…"*[36] If this is true, there is a need to re-envision the structure of our universe, where information occupies the ground floor when we consider its formation. This is mapped in Figure 4.2.

FIGURE 4.2: MAPPING THE INFORMATIONAL STRUCTURE OF THE UNIVERSE

Platforms/Systems		Mechanisms		Outcomes
Space-time	→	Classical Mechanics/ Relativity Theory	→	Determinism (Logic/certainty)
↓				
Quantum Domain	→	Quantum Mechanics	→	In-determinism (Choice)
↓				
Information Realm	→	Information Mechanics	→	Mind/consciousness (Awareness)

For those who have not previously viewed the Universe through an informational prism, this arrangement may appear strange, but it is supported by what can be deduced from scientific observation and offers explanatory power that has been lacking in previous models. In this chapter we have discussed what needs to be known to study and understand information. Now that we appreciate the nuts and bolts of information more fully, in Chapter 5 we will explore what we learn about the Universe by studying one of its informational outputs or products – life.

CHAPTER FIVE

BUILDING COMPLEXITY

"How did stupid atoms spontaneously write their own software?"

– Paul Davies[1]

We have previously noted that for information to be meaningful, it needs to be bundled with other information to provide context to discern meaning that builds knowledge. A number of different examples were listed in Table 3.2. This structured compilation of information creates complexity from which new and different entities can emerge. Our solar system is an example where matter and the natural laws of the Universe form a beautiful and elegant system. However, the most extraordinary complexity in our Universe is that which supports life and mind, so in this chapter, we shall be investigating the central reason that you can read and understand this page: Life!

5.1 – WHAT IS LIFE?

At university, I was taught that life is an emergent property of material entities. We did not know how or why; it was 'just so!' This is a very ancient discussion which has so far eluded a satisfying explanation, where there tend to be more questions than answers, so could we have been focusing on the wrong principles that might benefit from a fresh approach; in particular, that life is *more* than just a material entity? Sara Walker, Paul Davies and George Ellis argue that: "*... a complete mechanistic account of living matter will still fail to capture what it means to be living.*"[2] So is life physical, material or otherwise? Life certainly appears to be an entity that depends on the pre-existing material organism (the platform) in which it is found. Further, this platform must operate in space-time, so the platform at least is physical but the condition that we identify as 'life' would seem to be more than just this 'shell'; more like a dynamic, non-material information process. In 1999, Davies summarised this duopoly as: "*...life is a felicitous blend of hardware and software.*"[3]

The corollary to the question, "what is life?", is: "How did life begin?" In *The Algorithmic Origins of Life*, Walker and Davies point out that the age-old debate about which came first: the chicken or the egg, is really a discussion about whether biochemistry preceded the genetics or vice versa. Re-stated in information terms this becomes: did the information required for *chemical processing* seen in the metabolism of the cell or the *digital information* within its genetics take precedence? Walker and Davies write: "*In approaching this debate, a common source of confusion stems from the fact that molecules play three distinct roles: structural, informational and chemical. ... The chicken-or-egg problem, as traditionally posed, ... amounts to a debate of whether analogue or digital hardware came first.*" They point out that a discussion about the primacy of digital or analogue *hardware*, completely ignores the elephant in the room, which is the non-material information being processed by the *software*, as well as the software itself.[4,5]

5.2 – NON-LIFE INTO LIFE

Alternatively, what if life was viewed as an autonomous, complex, continuous, functional information process that is the result of an organism constantly making *internal*, information-driven, biochemical adjustments as it sustains and reproduces itself. By definition, life is dynamic or changing, so without the ability to continuously regulate, the organism can no longer maintain its essential functions and it soon dies. With the exception of viruses and some bacteria, living things operate as stand-alone units that run on internally stored information. The information processing can involve relatively simple routines such as are found in single-celled microbes or complex, multimodal processes as in mammals, involving multi-layered functions such as breathing, blood circulation and consciousness. Yet no matter the complexity, life itself is an all or nothing process and is either operational or absent, so to remain living, it must be continuous. To simultaneously achieve all these parameters, life requires a sophisticated hybrid mechanism that involves digital information processing capability running in parallel with analogue hardware, driven by non-material information programmes.

Walker and Davies conclude that the boundary that divides non-life from life lies at the point where "*information gains causal efficacy over the matter it is instantiated in.*"[4] In other words, the material entity (its physical body) is no longer principally controlled by its material composition or its physical surroundings but by its intrinsic (internal) information; the organism has gained self-autonomy from its environment (which is *external* control). When information melds with matter such that the combined entity can act independently of its environment, which is to say its internal information systems predominate over the external information that seeks to control it, we observe the phenomenon which we call 'life'. So, if life did evolve from non-life as Darwin supposed, the *transition* from non-life to life occurred when the information systems that formed its locus of control moved from its external environment to within its compositional fabric, such that it achieved self-autonomy. The question is how that might have happened. Conversely, when that entity loses all ability to influence its environment, it begins to decline and if prolonged it will experience death.

Walker and Davies also note that the information found in life is characteristically globally organised through a *distributed* information system which involves both analogue and digital information processing and displays top-down causation. So, life is an entity composed of informationally empowered matter; an integrated compilation of both material and non-material components that has self-determination over both its internal and external environments. In summary: "*...life may thus be identified when information gains top-down causal efficacy over the matter that instantiates it.*" The controlling information flow is therefore from global to local scales rather than local to global.[4]

This leads Walker and Davies to their next conundrum: "*...the 'hard problem of life' is the identification of the actual physical mechanism that permits information to gain causal purchase over matter. This view is not accommodated in our current approaches to physics.*"[6] And this is not for want of trying, so once again, a new approach is required. The significant question is: Is the process *wholly* operating within space-time where almost all the research has been focused, or could the quantum domain also be involved? Schwartz, Stapp and Beauregard note that the atoms that constitute our molecules and biochemistry operate within the quantum world, so it would seem likely that if the answer cannot be found in space-time, we may be operating, even if indirectly, on a Planck scale foundation in the quantum realm.[7] Thus, life might be an emergent process that runs on *quantum information* within a sufficiently complex physical organism (vehicle) that exists in space-time.

5.3 – INFORMATION IN LIFE

Many scientists are now comfortable with the idea that information is intrinsic to life. The debate is now shifting to its roles and sources. In 1990, Bernd-Olaf Küppers wrote: "*Ever since the elucidation of the molecular basis of living systems, we have known that all elementary processes of life are governed by information. Thus, information turns out to be a key concept in understanding living matter.*"[8]

Since Küppers penned this piece, our knowledge of the mechanics that operate the functions of the average cell has exploded. John Lennox observes that: "*... in molecular biology... proteins, flagella, cells, etc. are molecular machines. They may well be more than machines but, at the level of their information processing capability, they are certainly (digital processing) machines.*"[9] To put this into context, this is very different from non-biological systems that we find in nature. Rocks and water are also composed of information but no one would suggest that in their natural state they are machines and even less that they might be digital information processors.

So biological systems are informationally complex in ways that non-biological systems have never achieved. They not only possess information repositories that specify their makeup, they have the ability to process (convert) that information into functional outputs, along with overarching management systems, such as an epigenome, which is another layer of meta-information that sits above the genome and directs the activities of the cell.[10] As a result, there is a threshold or lower complexity boundary that defines the minimum information that is necessary to produce a viable cell that is able to exhibit life. This is an example of how information can also define boundaries. This concept was first introduced by biochemist, Michael Behe, who ignited a firestorm when he introduced the term: '*irreducible complexity*' in his 1996 book: *Darwin's Black Box*.[11] There was a lot of push back from other scientists, such as Jerry Coyne[12] but Stephen Jay Gould (1941-2002), the doyen of evolutionary biology acknowledged the need for irreducibly complex systems in 2002 when he wrote: "*Classical science, with its preferences for reduction to a few controlling factors of causality, was triumphantly successful for relatively simple systems like planetary motion and the periodic table of the elements. But irreducibly complex systems - that is, most of the interesting phenomena of biology, human society and history - cannot be so explained. We need new philosophies and models ...*"[13] By this statement, Gould is offering us a clue. In terms of their physics, planetary systems and inorganic chemistry are (informationally) relatively simple systems, that can be understood through a reductionist, bottom-up analysis but information systems that are irreducibly complex, which includes living systems, cannot be so explained and as indicated by Walker and Davies above, require top-down explanations. Davies had previously written in 1999: "*There seems to be an enigmatic circularity to life, a type of irreducible complexity...*"[14] If this is starting to appear intricate and involved, it is! As a consequence, the scope, depth and convolution of the myriad discoveries over the past half century have spawned the new discipline of 'Systems Biology'.

5.4 – SYSTEMS BIOLOGY

As our understanding of the physical complexity of biological systems has been growing, the mechanisms by which living things operate is giving rise to new questions. Professor Emeritus of Physiology at Oxford, Dennis Noble suggests: "*I think that as a gene-centric view of evolution, the modern synthesis has got causality in biology wrong. Genes ... if they're defined as DNA sequences, are purely passive. DNA on its own does absolutely nothing until activated by the rest of the system...*". DNA has usually been seen as an information storage unit, however life has turned out to be a dynamic information process, so "*a passive data base*" cannot be the whole story.[15] In fact, we have subsequently learnt that there is a multi-layered information system for all life that requires:

A. An active information management system that administers...

B. The cell's information repository or library for intracellular maintenance but also...

C. The routine external operations of the creature such as seeking food and locomotion;

D. It also drives the physical development of the body of the organism that forms the hardware of multicellular creatures as they develop from a single fertilised zygote;

E. Plus, internal cellular information also instigates and performs replication and...

F. Complex animals, such as mammals, may also have thought, social functions and language.

Traditionally, biochemical reactions have been studied in isolation from each other, to learn what each reaction does. This reductionist approach has led to an understanding of many of the processes occurring within cells but we now realise that there are tens of thousands of proteins and chemicals in every cell with many reactions occurring simultaneously in a microscopic chemical sea, tightly controlled by accelerator molecules called enzymes which speed them up or slow them down (in their relative absence). There is also the higher-level meta-information that constitutes the epigenome that operates through control molecules in the form of short strands of messenger-RNA (mRNA) that tweaks production lines through feedback loops. Further, the DNA information store in the nucleus of most cells is now realised to be only part of the story, with some of the information being disseminated through the cytoplasm as well, such as in the mitochondria, which are the cell's powerplants as well as other cytoplasmic production machines such as the ribosomes.[16] Noble asserts that this explosion in our understanding of the complexity of living systems has changed what we recognise as the minimally viable unit of life, from the genome to the cell.[17] So, whereas we used to try to understand living things at the level of the gene, the push now is to comprehend life at a cellular, tissue or organismal level which is called: 'Systems Biology'. The information dynamics involved in tens of thousands of messages being executed *simultaneously* within a cell is beyond imagination. Lennox suggests that this hyper-complexity is leading to a greater realisation of the inadequacy of a reductionist mindset in understanding the environment within the cell, that is better described by Systems Biology and the necessity of top-down management and hence analysis.[18] With all this in mind, it is now time to examine how life might have begun.

5.5 – THE ORIGIN OF LIFE

The problem with Systems Biology is: where does the initiation and ignition of these information systems come from, in the light of our previous discussion regarding the need for both analogue and digital systems being integrated in this process?[19] Once again, this conundrum goes back to antiquity but since Charles Darwin published his *Origin of Species* in 1859, Darwinian evolution and its various subsequent iterations over the past 150 years has been generally accepted by mainstream science as *the* answer. However, the accelerating analysis of information over the past thirty years has led to some questioning its validity.

Microbiologists and molecular geneticists, Professor Christoph Adami and Dr Thomas LaBar write: "*One of the defining characteristics of life (perhaps the defining characteristic) is that life encodes information. … but evolution cannot explain where the first bits came from. … The 'infinite monkey theorem' posits that a million monkeys typing on a million keyboards, if given enough time (and typing randomly) could ultimately type out all of Shakespeare's works. However, the theorem is misleading, as even correctly typing out the first 30 characters of Hamlet's soliloquy ('To be or not to be …') cannot occur during the time our universe has been around (about 4.36 x 10^{17} seconds), as Hamlet's 30-mer* [initial characters] *is one in about 3 x 10^{42}. Using biased typewriters will not allow the monkeys to finish either, as it is only accelerating the search by a factor E ≈ 46,700.*"[20]

Another way of expressing this is that Darwinian evolution is analogous to starting with the English letter 'a' or 'I' and letter by letter writing Shakespeare's *Hamlet* (his longest play), with the proviso that every letter adds function (which constitutes *meaning* in the context of the play). If any additional letter fails to make sense (is dysfunctional in the case of an organism), the project is lost and must be started again. Bear in mind that Hamlet has approximately 200 *thousand* characters (including spaces)[21] while the human genome has over three *billion* characters. The task is monumental!

5.6 – SELF-ORGANISING SYSTEMS

Richard Dawkins' response has been that although this does appear to be an exercise in scaling '*Mount Improbable*',[22] it can be achieved by starting with simple systems, such as small bacteria and the complexity can then build to form complicated organisms such as sea sponges and eventually progressing to complex entities such as fish and larger creatures. The problem is that the implications of the Infinite Monkey Theorem means that even the simplest bacterium is not just improbable but decidedly impossible![23] As a result, Stuart Kaufman has taken a different approach and proposed *self-organising systems*, using computers to model them, to speed up the processes, as the time periods required are immense.[24] The problem is that the computers that demonstrate the process are not self-organising but intelligently designed. However, even without a computer to illustrate the process, any system that is able to organise itself, inevitably is constructed from further meaningful information that must also be explained away before we can get to the appearance of the first organism.[25] Perhaps the most difficult problem though is how these essentially hardware-based systems converted to living software-based systems. Paul Davies writes: "*The theory of self-organisation as yet gives no clue how the transition is to be made between spontaneous, or self-induced, organisation - which in even the most elaborate non-biological examples still involves relatively simple structures - and the highly complex, information-based, genetic organisation of living things. ... It is not enough to know how these giant molecules arose or started to interact. We also need to know how the system software came into existence. ... This is not merely a matter of adding an extra layer of complexity, it is about a fundamental transformation in the very nature of the system.*"[26]

This has significant implications for Darwinian Evolution. Küppers' comments regarding the generation of new information in Chapter 4, that to generate a piece of information of a given complexity requires another piece of information of the same or greater complexity, has important implications when we consider origins; both for the laws governing our inanimate universe but also for the origin of life. Each requires vast amounts of new information, but where could it have come from? Regarding life in particular, Professor Emeritus of Biology of the University of Vienna, Gerd Müller writes: "*... natural selection, which can account for the modification and loss of parts ... has no innovative capacity: it eliminates or maintains what* [already] *exists.*"[27] Even after the first organism appeared, however that may have occurred, there remains another problem. As Müller says, natural selection is an excellent mechanism for sifting out the inefficient or broken iterations of inheritance, but random mutation is not nearly a powerful enough mechanism to generate the code required to produce even a tiny fraction of the minimal entry-level, meaningful information necessary for life. Michael Behe showed that single mutations are common, felicitous double mutations are rare and co-ordinated triple mutations are nigh impossible; so to achieve the thousands, verging on tens of thousands, of harmonised mutations necessary is utterly impossible within the timeframes available since the formation of the Universe, unless there is a guiding mechanism to expedite the process.[28] Darwinists will argue that these tens of thousands of mutations do not have to happen simultaneously but without a blueprint to tell the organism ahead of time which mutations will be beneficial after another 9000 mutations, means it has no reason to conserve the other 9999 mutations that would have been useful, while awaiting the last one to make the intracellular machine functional. So, the blind random mutations supplied through Darwinian Evolution cannot provide the sought-after solution to the problem of generating new meaningful biological information.

5.7 – INVOKING AN ALGORITHMIC ORIGIN

Yet another suggestion has been that as the Universe is thought to have been lifeless at its inception, could a law of the universe have initiated a relatively simple algorithm to create life from non-life? John Lennox points out that this would require a "*...law that could produce the specified randomness of the coding of information on the genome. Hubert Yockey, together with Dean Overman, wrote: '... the physical laws are not good candidates for the source of life's origin. By definition, a law of nature is a very short algorithm... Life requires much more information than contained in these laws. The genetic information contained in even the smallest living organism is much larger than the information contained in the laws of physics and chemistry.*'"[29] Yet Bernd-Olaf Küppers has previously told us that the progenitor entity's information would need to be of the same complexity or greater than the new entity, so neither can the Laws of the Universe generate the biological information required for life.

No other satisfactory scientific solutions are currently considered likely explanations either. John Lennox writes: "*Yet surely,* what *organisms* actually are - *the very nature of the complexity of organisms, the nature and extreme sophistication of their protein machinery, the coding they contain, their capacity to use programmed information, etc. is piling up evidence that there is a mind behind life, even though we have little or no idea how life itself came to be. It is, after all, a plain fact that the only known source of information of the type we have been studying is mind.*"[30] The problem is that in the past, we have looked to reductionism to solve this conundrum. Reductionism can investigate bottom-up causation, where the information content of the construction is simply the sum of its parts. The establishment of an integrated system requires top-down causation which is information dependent and must be pre-specified by information from outside the system, which is hence adding new external information; so a top-down system is therefore *more than* simply the sum of its parts.[31]

Thus, Information Mechanics demonstrates why we are still waiting for the explanation of how materialism could explain the appearance of the first cell. Lennox concludes: "*The mind-boggling levels of informational sophistication at the epigenetic level* [make] *it.. clear … that reductionist attempts to explain biogenesis do not work and the modern (neo-Darwinian) synthesis has failed – not in the sense that they have not taught us useful science, but in the sense that they have not reached their stated goal of giving a satisfactory account of the genesis of the evolution of life. … Unsurprisingly, there is also a marked reluctance to grasp the inability of unguided natural processes to produce 'language' encoding of DNA.*"[32]

So, as bottom-up causation does not seem capable of explaining the origin of life, it would seem that we need to look for top-down solutions. This will require a higher-level view of integrated biology; yet where to start? John Lennox observes: "*... the only source* we know *of the kind of language type information that we have found in the living cell is mind.*"[33] So our next port of call is the mind.

CHAPTER SIX

THE MIND CONUNDRUM

"Nothing makes sense in biology, except in light of information."

– Bernd-Olaf Küppers[1]

The mind has so far proved impervious to external direct interrogation despite the search having been long! Two thousand years ago, Seneca the Younger (c.4 BCE-65 CE) wrote: *"Speech is the index of the mind."* Even then, Seneca seems to have appreciated that to know someone's thoughts requires the subject to reveal them, such as through speech. We can take x-rays, CT scans, MRIs, PET scans, EEGs, polygraphs, you name it but there is no way we can identify a person's thoughts unless they tell us. EEGs, PET scans and functional MRIs can witness the brain in action but cannot identify *what* the person is thinking.

6.1 – UNDERSTANDING THE HUMAN MIND

Because the brain and mind are so poorly understood, there are many theories. In naturalism, mind can be non-material whereas in materialism, the mind is merely the output or product of a functioning brain.[2] However, if man is simply the end result of material particles in motion, what reason have we for believing that all those molecules moving deterministically according to the dictates of the laws of the Universe, can have any meaning, let alone convey that our thoughts are even valid? There is none at all! If that was the case, then we can end the book here, because the only logical conclusion is that everything is meaningless. Yet, from their lived experience, most people believe that their thoughts are valid, so it does not seem unreasonable to suppose that mind is more than just deterministic-molecules-in-motion.

Instead, scientists are increasingly becoming aware that we all have both consciousness and self-awareness. Consciousness is an awareness of our surroundings that arises spontaneously from within us and is such an elusive concept that most of us rarely contemplate it, but we use it all our waking hours and it needs an explanation. Yet science is currently at a loss to even know where to start. Oliver Sachs, Professor of Neurology, lamented that despite all our advances in neuroscience *"'...neurological activity and psychic activity still seem utterly different,' even 'incommensurable' in kind."*[3] Consciousness is like an operating system on a computer; it is a platform on which other useful programmes, such as thoughts and feelings, run. Without this software package there is no useful output or productivity from the computer. However, the hardware system [our physical brain] by itself yields nothing either. It is the integrated hardware and software package when combined into one active system that is productive - once again, layer upon layer of integrated information!

6.2 – SELF AWARENESS

In human beings, self-awareness is the complementary twin to consciousness. Consciousness allows us to survey our exterior environment in space-time, whereas self-awareness analyses our interior realm of thoughts in Mindspace, which would appear to be reliant upon the quantum domain which will be explored in Chapter 7. No one doubts that self-awareness is a trait possessed by all normal adult human beings. However, are chimpanzees, which are generally considered the next most intelligent species after man, self-aware? The evidence would suggest that they are not. They do not wear clothes or apparel for modesty (whereas almost every human people group does). Neither do they appear to reflect on their own existence or mortality; no chimps have been observed to practise funerial rights or create memorials to the dead as human beings do. Although primates can be observed to be thinking and able to perform simple problem-solving exercises, there is no evidence that they possess self-awareness. Without an ability to self-reflect, animals are not considered morally responsible for their actions; human beings can be tried for murder but sharks, crocodiles and grizzly bears cannot, as they are not moral creatures; they have no concept of the consequences of their actions and hence a sense of right or wrong.

6.3 – THEORY OF MIND

Robin Dunbar, Professor of Evolutionary Anthropology at the University of Oxford observes that despite all the fuss being made about cultural traditions in chimpanzees, orangutans and whales, only one species has built the Great Wall of China, put men on the moon, thought up Pythagoras' theorem, conceived the existence of life after death, written the plays of Shakespeare and composed the music of Johann Sebastian Bach. *"In short, it is in humans' capacity for culture, our capacity to live in a world constructed by ideas, that we really differ from the other apes..."* Dunbar cites the concept known as 'theory of mind' - the capacity to imagine other individuals 'mind states' - as a crucial difference between humans and other animals. *"It is widely accepted now that most higher animals have knowledge of their own mind states. In terms of philosophy of mind, this is what is usually referred to as first-order intentionality. 'Intentionality' is a general term covering those mind states that involve beliefs. They are evidenced by the use of words like believe [that something is the case], know, suppose, intend, want, desire, wish, etc. They imply that we have a certain level of awareness about the contents of our minds, that we are aware that we believe... that something is the case."*[4]

First-order intentionality is the world inhabited by very young children. It is not self-consciousness but is certainly a form of consciousness that is shared with birds and mammals. However, around four years of age, children develop the capacity to understand the minds of others as well. This capacity is referred to as 'theory of mind'. Once they have acquired the capacity for theory of mind, they are able to engage in imaginative play, that a doll can drink tea from a teacup, even though the cup is empty, and they can lie with conviction.[4]

"With theory of mind in the mental toolkit of young children, they become good psychologists - they understand the mind behind the behaviour and begin to understand how false information can mislead that mind. They are in what philosophers of mind would call second-order intentionality - they believe that someone else believes [that something is the case]. *Before this, they have always assumed that everyone else shares the same knowledge of the world as they do."* Dunbar concludes: *"There is no strong evidence to suggest that any species of animal other than humans has theory of mind."*[4] Yet in human beings, babies have first-order intentionality, 5-year-olds have second-order and normal adults scale in at fifth-order.[4]

The fact that birds, mammals and human beings have consciousness but that only human beings possess self-awareness suggests that there is a category difference between human beings and other living creatures, rather than this being the different ends of a spectrum, since there is a chasm of gigantic proportions between the cognitive abilities of chimpanzees and adult human beings.[5]

So, to function normally as human beings, we need both consciousness and self-awareness. Even in 1983, Joseph Levine expressed the opinion that there is an 'explanatory gap' between materialism and mind.[6] David Chalmers in: *The Conscious Mind* writes that the mental faculties represent a 'non-reductive primitive' by which he means they are a fundamental building block of the Universe in the same category as matter/energy or space and time.[7] This is supported by the observation that meaning can only be created, conferred or appreciated by mind. Meaning is never comprehended by material objects in isolation. As a result, we need to view our mind and brain as a co-ordinated unit: the *mind-brain complex*, which is a fusion of both the physical brain and the non-material mind that can now bridge the explanatory gap.

6.4 – THE QUANTUM MIND

It has been known for a long time that the mind-brain complex (MBC) is an information processor. Information is input into the MBC in one of two ways: it can enter through our senses of sight, smell, hearing, taste and touch or it can be generated within the mind de novo. Through our senses we sample our external space-time environment and through our mind we manage our internal environment in Mindspace. The mystery has always been how? It now appears that our mind may be a 'quantum computer' that can harness or interrogate the quantum domain from within. Through analysis of MRI scans of the human brain, physicist, Dr Christian Kerskens of Trinity College, Dublin has detected entanglement of nuclear proton spins which requires quantum involvement. He writes: *"Because these brain functions were also correlated to short-term memory performance and conscious awareness, it is likely that those quantum processes are an important part of our cognitive and conscious brain functions."*[8]

Kerskens' results are suggestive that our MBC is using quantum processing but how? In 2014, anaesthetist, Stuart Hameroff and mathematical physicist, Roger Penrose proposed that quantum processing might be occurring within the microtubules in neurons, which determines axonal firing. This quantum processing may be instrumental in consciousness and thought.[9] It is proposed that the quantum wave states are isolated within the tiny microtubules but that these sensitive quantum states collapse as they leave the microtubules and interact with the surrounding nerve cells, as they interact with space-time. The intriguing observation is that these microtubules also form the cytoskeleton in all complex cells and are therefore found even in an amoeba.[10] Might this be the mechanism by which single cells are also able to process information, such as responding to light and food? In a recent presentation, Stuart Hameroff suggests that this is so. Plants and amoebae lack human mental abilities as the processing power of microtubules is proportional to the density of their bundling that affects their resonance capabilities. In plants and one-celled organisms, microtubules number in the thousands per cell but in human nerve cells they are packaged in tens of millions.[11]

6.5 – MODELS OF THE MIND

However, Kerskens, Penrose and Hameroff are not the only ones researching this field. Over the past 70 years, many researchers have weighed in on this elusive subject. Some doubted its existence such as Daniel Dennett (1942-2024)[12] and Viktor Stenger (1935-2014),[13] some like Eugene Wigner (1902-1995),[14] considered it was necessary but could not elucidate a detailed model, while others have very detailed models with growing bodies of experimental evidence, such as the Penrose-Hameroff Orch-OR Model,[15] Jibu and Yasue's Quantum Brain Dynamics[16] as well as CNET[17] to name a few. All these models imply a quantum mechanism for consciousness and brain function. It is possible that more than one of these mechanisms might be operating in different parts of our brain simultaneously as there are multiple roles to be performed and even if Occam's razor suggests a single solution is superior to multiple ones, redundancy is not an uncommon feature in the human body.[18] The problem is discerning which.

Then, in 2024, three papers were published that presented evidence for quantum involvement in consciousness. Nishiyama et al begin with the Quantum Brain Dynamics Model which: "*...adopts photons and water as main actors ...[The] Brain is a mixed system of classical neurons and* *glial cells on the one hand, and quantum degrees of freedom of photons and water molecules on the other hand. The latter obey the laws of quantum field theory...*" They continue: "*Microtubules...form nonlinear optical devices akin to lasers, creating an electromagnetic field inside their hollow core in a process termed 'super radiance'. The optical computing proposed to occur in these microtubule networks was* [considered] *to provide the basis for cognitive functions.... Tubulin proteins are envisaged to interact with their neighbouring tubulin molecules in a microtubule and by doing so perform quantum computations using entanglement effects... [A] super-radiance solution in holographic brain theory...*" was also derived mathematically. Then they calculated the holographic memory capacity of the brain's neocortex which has a potential many magnitudes greater than classical theory.[19]

Two months later Babcock et al reported that after stimulating microtubule networks with UV light, they also observed "*...superradiant behaviour in the fluorescence...[of] tubulin structures...*" at body temperature.[20] Then several months later, Khan et al described how "*anaesthetic gases ... bind to cytoskeletal microtubules and dampen their quantum optical effects, potentially contributing to causing unconsciousness.*" They conclude that "*...quantum dipole vibrations in microtubules inside neurons mediate consciousness, functional connectivity and regulate neuronal firing.*"[21]

6.6 – QUANTUM BIOLOGY

For those readers who are not familiar with quantum biology, this probably sounds like science fiction gobbledygook. Restated, the ultra-fine hollow microtubules that are inside our nerve cells and composed of multiple copies of a special protein called tubulin, have been shown to fluoresce and act as optical computers through quantum effects. Anaesthetic agents disrupt the microtubular oscillations, which induces unconsciousness, suggesting that the microtubules are important for consciousness.[22] Furthermore, the brain can process quantum holograms which can store vast amounts of information which is a possible explanation for memory.

The take home message from this research is that since the binding of anaesthetic agents to the microtubules induces anaesthesia (unconsciousness) and wakefulness is re-established by removing these agents, microtubules in the cerebral cortex of our MBC appear to be integral in conferring consciousness. The mechanism appears to combine quantum oscillations or resonance in benzene rings in the amino acid, tryptophan, in the tubulin sub-units within the microtubules.[11] However, that still leaves the question: what controls the activation of the microtubules - which at present is unknown, although Hameroff notes that consciousness appears to be at the interface between space-time and the quantum realm and may require an interplay between the two.

6.7 – THE RESONANT MIND

For those who are struggling with the above explanation, this section on detailed mechanisms can be omitted without compromising your understanding of the way that we understand consciousness.

Hameroff goes on to explain that within the amino acids, "*... hexagonal, 6-carbon benzyl/phenyl rings share 3 de-located 'pi orbital electrons' among the six carbon [atoms] forming 'pi resonance clouds' conducive to quantum effects [that] support quantum electric and magnetic dipole oscillations, excitations, spin transfer, photons, fluorescence, phosphorescence, sub-radiance and super-radiance. These are quantum effects,...even at ambient temperatures.*" The aromatic rings when in a geometric array form 'pi resonance groups' that can oscillate like little dipoles (one bar magnets) at a frequency in the order of terahertz (10^{12} Hz), which is the source of the oscillations. Herbert Fröhlich (1905-1991) proposed this phenomenon, termed Fröhlich coherence, in 1967. Hameroff reports that the greater the bundling of the microtubules, the faster the oscillatory speed of the system, while anaesthetics block or jumble these dipole oscillations. Human beings experience around 1 billion events (10^{8}) per second while plants experience one event every few seconds.[11] Cognitive psychologist, Professor Donald Hoffman describes how a similar interference effect can be achieved by using a focussed magnetic field, such as is used in Transcranial Magnetic Stimulation (TMS). Colour vision temporarily fades to grey under magnetic stimulation and slowly returns after the TMS is turned off.[23] This is understandable if consciousness is dependent on highly co-ordinated electro-magnetic oscillations that are disrupted by the TMS.

In 2000, Max Tegmark estimated that the decoherence time in microtubules would be too fast to allow quantum coherence events to occur, based on the macular event times measurable on EEGs,[24] but Hameroff queried Tegmark's methodology and points out that the internal oscillation frequencies are many orders of magnitude faster than the electrical activity measured by an EEG that is thought to record axonal firing in space-time rather than the internal microtubules operating in the quantum realm. Penrose and Hameroff suggest that a conscious event results every time that decoherence occurs, which means that if the frequency of microtubular conscious events is faster than the speed of decoherence, then there should be a continuous stream of consciousness, whereas if the decoherence time scale is faster, as Tegmark suggests, there should be an absence of continuous consciousness. Thus, the threshold for continuous consciousness in living organisms is determined by a microtubular density that provides an oscillatory speed greater than the speed of decoherence which Hameroff calculates to be between 10^4 and 10^6 cycles per second.[25]

The rapidly repeating decoherence events within the microtubules described by Penrose and Hameroff can be understood as a superposition in the tubules collapsing every 10 millionth of a second which is to say that information is transferred from the quantum domain to space-time neurons 10 million times per second, which Hameroff maintains is 100 to 10,000 times faster than the decoherence rate. The corollary of this is that organisms that have insufficient microtubular bundles to oscillate this fast (such as bacteria and plants), would not experience continuous consciousness but would experience discrete events that may be useful for (sub-conscious) information processing.[11] So, consciousness appears to derive from quantum processing effects that are auspiced by intrinsic quantum-mediated information events within the tubules inside space-time neurones.

6.8 – MIND FIELDS

Neuropsychologist, Brian Whitworth suggests that as bacteria also exhibit quantum effects, they can be used to help understand quantum effects in other life forms. He writes: *"Studies show that bacteria chromophores vibrate in step to give quantum beats (Engel, 2007) that allow their electromagnetic fields to entangle (Maiuri, 2018)... However, identical matter entities whose quantum fields overlap will entangle (Lo, Franco & Compagno, 2016)...*[so] *Molecules vibrating in exact synchrony maintain a coherence that is usually lost in the molecular bustle of a normal cell. The timing of the synchrony must be almost perfect to produce this quantum effect,...* [M]*icrotubules allow synchronous oscillations that give strong Fröhlich coherence at room temperatures (Samsonovich et al., 1992). If the cell structure oscillates synchronously, molecules in it will do the same, allowing them to superpose, cohere and entangle to act as one. That microtubules can apply quantum effects at cell time scales has led to their study in biological puzzles like smell, protein folding, ion channels, and bird navigation (Gauger, 2011).*

Evidence that neural synchronies correlate with consciousness include that they:... Don't depend on firing rate. Studies of the visual cortex find that the: "... selection of responses for further processing is associated with enhanced synchronisation rather than increased firing." (Singer, 1999 p62).... If the brain is a network of oscillators, a nerve is more like a Wi-Fi device than a transistor chip. Brain areas aren't just wired together, they resonate together, suggesting that consciousness arises in the brain's electromagnetic field. Conscious electromagnetic field information (CEMI) theory suggests that: "... the brain's EM (electromagnetic) field is the physical substrate of consciousness." (McFadden, 2020) p5.... McFadden argues that data processing can't integrate information but an electromagnetic field can... That the mind is in the [EM] *field solves the mind-body problem at a stroke, as then consciousness is unified because: "... EM fields are always unified, there is only ever one EM field in the brain." (Ibid, p6)... We download data from the brain's electromagnetic field as we download songs from a Wi-Fi field..."*[26]

This model suggests a compound mechanism for consciousness (the operating system) and (from the preceding discussion) it would appear that it also accounts for conscious thought and memory too. Hameroff suggests that degeneration of the microtubules (resulting in dislodgement of the Tau protein subunits) is a likely mechanism for Alzheimer's Dementia,[11] where deterioration of short-term memory recall is an early effect compared with logical thought which follows later, while the clouding of consciousness is a late effect. So, it is possible that the laying down of *short-term* memory also requires microtubular processing for space-time storage.[27] In contrast, compromise of *long-term* memory in

Alzheimer's is relatively late, so that is proposed as a quantum domain (holographic) storage phenomenon.

6.9 – SLEEP

A logical extrapolation of these discoveries is that, if the decoherence rate is faster than the tubulin oscillation rate (as in an amoeba), then the tubulins can act as an information system but without consciousness. This suggests a model for sleep, which might be induced by the MBC *intentionally* slowing down the oscillation rate to below the decoherence rate, so that consciousness would be suspended as in deep sleep, which is also called slow-wave sleep. To wake up again, one only has to increase the tubulin oscillation rate above the decoherence threshold once more. This model for sleep is consistent with the observation that the unconsciousness induced by anaesthesia is different to natural sleep, however more research is needed to find a mechanism to orchestrate this oscillatory control. Light sleep would then be the phase around the decoherence rate and seems to be the 'neutral gear' that the brain uses to switch between deep sleep, REM sleep and consciousness. Rapid eye movement (REM) sleep is associated with dreaming and is more complicated as the physical space-time brain is unconscious, yet the quantum mind is active. During dreams, the MBC temporarily disconnects itself from the physical body, as if one is paralysed, which requires more mechanisms that are not yet understood. Concussion and permanent brain damage relate to *physical* disruption of the harmonic tubulin resonance systems following trauma and seems to be proportional to the degree of injury.

But if our cells are quantum processors, where does this leave materialism, because quantum processes are not deterministic? This includes our neurones; if neurones are not deterministic then neither is the MBC. Professor Emeritus of Philosophy and Law, Thomas Nagle observes: *"But if the mental is not itself merely physical, it cannot be fully explained by physical science… Materialism requires reductionism; therefore the failure of reductionism requires an alternative to materialism."*[28]

6.10 – SHIFTING OUR THINKING

Schwartz, Stapp and Beauregard also conclude that, as at the fundamental level, all physical systems depend upon quantum processing for their functionality, so the neurons in the human brain must also.[29] This means that at their foundation, our brains are quantum objects, which is not the way most scientists are taught at universities nor physicians in medical schools. Lennox observes: "*It seems remarkable that, although QM is nearly a century old, relatively few people seem to grasp that it has demolished the materialist worldview with its mechanistic determinism.*"[30]

Now, in Newton's classical mechanics, the observer and mind have no effect on interactions but John Lennox writes: "*...in the quantum world, ... [o]bservation becomes an interplay of mind and matter and involves non-material causation. The universe revealed by QM is not a materialistic universe. It involves psycho-physical effects. ...and, as such, it presents us with a new perception of reality.*"[31] A significant example of this is the way that thought drives the development of the human brain, with the differentiation of stem cells to produce new neuronal pathways based on a 'use-it-or-lose-it' principle. For example, if we are learning a new skill such as tennis or golf, repetition of the thought processes involved, amplifies the growth in these nerve tracts; active thought that results in learning, changes our *physical* brain structure![32] This jarring rethink on the functioning of the Universe may need a period of reflection for many of us but a significant number of quantum physicists agree. Henry Stapp wrote: "*The classical theory presumes that all aspects of nature can be explained purely in terms of the action of matter upon matter. But the quantum world differs in a fundamental way from that core precept of materialism. This huge structural difference in the real (that is, quantum) world between the matter-matter interaction and the matter-mind interaction makes manifest the extreme naïveté of trying to comprehend the connection between mind and matter within a materialistic framework.*"[33] Floridi points out that this is well demonstrated in classical deduction where the mind introduces new intermediate thoughts as it works from one position to another. However, on reaching the conclusion, the intermediate steps are dispensed with, leaving no material trace at the end of the process but the non-material elements were required to reach the end point.[34]

These ideas are corroborated by cognitive neuroscientist and Nobel laureate Roger Sperry (1913-1994) who wrote: "*The laws of biophysics and biochemistry are not adequate to account for the cognitive sequencing of a train of thought.*"[35] Thus, mind is independent of matter. Fellow neurobiologist and Nobel laureate, George Wald (1906-1997) went even further when he wrote: "*Consciousness seems to me to be wholly impervious to science. It does not lie as an indigestible element within science, but just the opposite: science is the highly digestible element within consciousness ... Mind, rather than emerging as a late outgrowth in the evolution of life, has existed always as the matrix, the source and condition of physical reality - that the stuff of which physical reality is composed is mind-stuff.*"[36]

6.11 – THE MIND-BRAIN COMPLEX

A century of detailed research into the mind-brain complex has not been able to build up a functional model from finer and finer analysis; the bottom-up approach has failed.[37] Instead, it would appear that a top-down approach is necessary as Lennox has previously suggested. A proposed model of the human mind-brain complex is illustrated in Figure 6.1. In Tier 1, we live our physical existence in space-time, which corresponds to the information describing the hardware (which neuroscientists sometimes refer to as 'wet-ware'). This is our physical brain and is a *vehicle* which hosts the higher functions of the human mind. This is the province of neurology and neuroscience. We experience this and feel we understand it but there is more. Tier 2 is the non-material software that runs on the physical platform of our brain. This is analogous to the operating system within our computers but unlike computers it provides us with consciousness, active thought and 'mind' and appears to operate holographically, an all or nothing phenomenon. In the 1920's, US neuropsychologist, Karl Lashley (1890-1958) discovered that after training rats to negotiate a maze to find food, it did not matter which part of their brain he removed, it did not erase their memory of how to negotiate the maze successfully, although it did affect the acquisition of new knowledge. This suggests that the memories were stored holographically, which is a quantum effect. This was confirmed by Karl Pribram in the 1960's.[38,39] Tier 2 is also involved with the processing of information and knowledge and is studied in psychiatry and psychology. Tier 3 involves self-awareness which is like an application programme that runs on the underlying physical hardware and operating system of consciousness. This processes meta-information and allows self-reflection, leading to understanding and wisdom. It is often seen as: '*the self*'. In this text it is referred to as the 'human spirit' and is the province of philosophy and theology.

FIGURE 6.1: THE HUMAN MIND-BRAIN COMPLEX

Tier	Component	Function	Output	Entity	Discipline
III	Self-awareness 'App'(programme)	Self-reflection	Understanding/ Wisdom	Spirit 'Self'	Philosophy/ Theology
	+				
II	Non-material 'Software'	Consciousness/ Thought	Information/ Knowledge	Mind	Psychiatry/ Psychology
	+				
I	Physical 'Hardware'	Material platform	Physical signals/ Data	Brain	Neurology/ Neuroscience

So, if the quantum domain underlies space-time as Stapp and others suggest above, our brain is a holographic quantum processor and every human being is an embodied (quantum) mind; we are firstly mind/ spirit and secondarily body. Further, it would seem that the evidence is pointing to microtubular quantum processing mechanisms that are not only present in *your* brain but that all life is a compilation of classical and quantum parts, even the humble amoeba and your decorative rose bush! It would appear that all life is quantum-enhanced via quantum information. So, it is time to recognise and accept the reality of *quantum biology*. Everything with microtubules, cilia or a cytoskeleton is likely to be quantum augmented, which makes the behaviour of insects such as ants and bees less surprising also.

6.12 – MONISM OR DUALISM

Since the Greek philosopher Plato (c.427-347 BC), there has been argument over whether the human mind is merely material, with an emergent thought capacity that is generated by the physical operations within the brain alone or the blending of both material and non-material elements during construction. This is the monist versus dualist debate. Before we begin however, we need to define what is intended by the terms monism and dualism. We have seen in the previous section that the human brain has both physical components and non-physical elements which are both essential for the proper processing of human mind. Removing either means the person can no longer function as a normal human being. To clarify this, consider whether your computer is monist or dualist. To describe it as *merely* hardware or *only* software is nonsense. Without both components it is not a functional computer. At its core, the question of monism versus dualism depends on the frame of reference. To be a human being requires the mind-brain complex to operate as one single entity (arguably monist) but if we drill down to the physical brain, it is demonstrably different from the non-material thinking and consciousness/self-awareness that constitutes mind (apparently dualist)! So, to describe a single (monist) or combined (dualist) model depends on the frame of reference and the way the terms are defined.

6.13 – INTERPRETATION OF MEANING

In Chapter 3 we saw that bundled information can convey meaning, however, there are very few entities in the Universe that can subsequently extract or understand that meaning. Floridi points out that only *"informational agents"* can *interpret* data and even then, only within an appropriate *"level of abstraction"* (frame of reference) that must be supplied by accompanying data.[40] At present, these informational agents can be: appropriately programmed machines (such as computers) or living creatures with complex brains. Floridi notes though, that regarding the *understanding* of meaningful information: *"As far as we know, we are the only semantic engines in the universe."*[41] There is therefore a category difference between the operation of the human mind and everything else that is currently known to science.

CHAPTER SEVEN

CLASSIFYING INFORMATION

"One doesn't discover new lands without consenting to lose sight of the shore..."
– Andre Gibe (1869-1951)[1]

We have now covered quite a number of disparate information-containing entities in the Universe but the problem has been to link the different pieces systematically. In this chapter we begin to draw the threads together to unite them into one model.

7.1 – DIFFERENT DESCRIPTIONS OF INFORMATION

Historically, a number of different ways have been developed to describe or classify information but none has been able to adequately account for the bulk of the information we encounter on a daily basis. In Chapter 1 we discussed *Shannon Information,* also called the *Mathematical Theory of Communication (MTC)* that measures the volume of information transferred and is hence quantitative. This describes the *throughput* but provides no understanding of the *content,* the message itself, that is being transferred. In contrast, *semantic information* is meaningful or descriptive, which is qualitative, but resists compression and hence digitisation which has frustrated mathematicians and scientists who like to build mathematical models to further their understanding. In an attempt to break this impasse, *Algorithmic Information Theory (AIT)* was devised by Andrey Kolmogorov and Gregory Chaitin to try to measure the content of a message by calculating the *complexity* of the communication, however it still does not address the *meaning* within the message.[2] Each of these have been useful in certain circumstances but none has provided an overarching model to describe or manage information in general. Shannon information, which is algorithmic can be compressed, so Paul Davies writes that mathematically it *"contains very little information"*, whereas meaningful information, which is non-algorithmic and hence incompressible is *"information-rich"*.[3] This makes sense as Shannon information describes the vehicles in space-time, which are less complex to specify than the (often hidden) meaningful information content that they contain. To integrate these significantly differing categories, it would seem that a new approach is needed. We have previously seen that from an information perspective, the Universe can be viewed as an amalgam of material and non-material entities and we will see that they can behave in one of four different ways, suggesting that there are four different forms or domains of information. Each is now examined separately.

7.2 – SPACE-TIME

We began investigating space-time when examining computer hardware and from the physical universe we derive material science which has been serving us well for four centuries. Figure 7.1 maps some features within the space-time domain. It illustrates the progression from fundamental particles that form atoms, to molecules and then substances that can be either organic, that underpin life, or inorganic, that make up inanimate entities, such as computer hardware. This has been extensively studied by science over centuries and is thought to be quite well understood.

FIGURE 7.1 – MAPPING SPACE-TIME

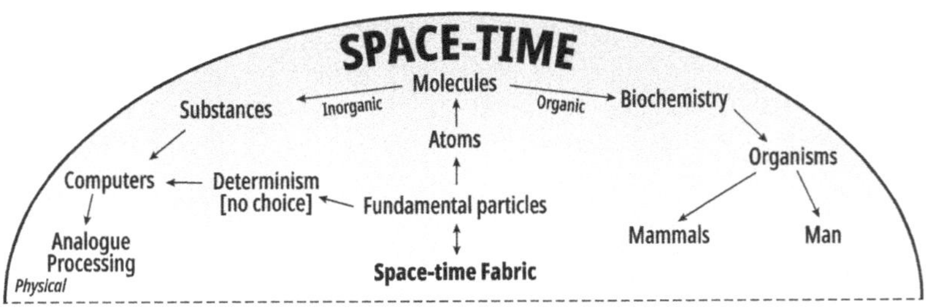

Underlying our familiar physical universe is an invisible entity that physicists call the space-time fabric. We cannot detect it directly, but we infer that it is there because objects called 'phantom particles' materialise out of the empty vacuum of vacant space and then vanish back into it without a trace. This space-time fabric appears to support our material universe by undergirding the fundamental particles that form atoms.[4]

Space-time contains both matter and energy but as discussed in Chapter 2, Einstein's Special Theory of Relativity shows that matter and energy are just different forms of the same thing ($E=mc^2$), so, there are only two different fundamental entities in the material universe – matter/energy that does things and information that tells it what to do.[5] It is significant to note that space-time is deterministic, so computers are too. However, we have already seen that computers are not *wholly* material, as non-material software is integral to their functioning, so next we need to examine Cyberspace.

7.3 – CYBERSPACE

The second domain is Cyberspace, wherever and whatever that is! Although this term has appeared in literature for over thirty years, no one has given a definitive explanation to account for its existence, further than the notion that it is a non-material realm involving information processing.[6] However, in 1968, Karl Popper introduced the concept of such a domain as one of his '*Three Worlds*'.[7] Then, in 1994, Roger Penrose further developed the concept, using the description: "*the Platonic world of mathematical forms. There, we find the natural numbers 0, 1, 2, 3,..., and the algebra of complex numbers.*"[8] However, having described the physical world of space-time and the mental world of thought, Penrose then goes on to ask: "*What right do we have to say that the Platonic world is actually a 'world', that can 'exist' in the same kind of sense in which the other two worlds exist? It may well seem to the reader to be just a rag-bag of abstract concepts that mathematicians have come up with from time to time. Yet its existence rests on the profound, timeless, and universal nature of these concepts, and on the fact that their laws are independent of those who discover them. The rag-bag - if indeed that is what it is - was not of our creation. The natural numbers were there before there were human beings, or indeed any other creature here on earth, and they will remain after all life has perished. ... Also in this world are Maxwell's electromagnetic equations as well as Einstein's gravitational equations... There is the mystery of why such precise and profoundly mathematical laws play such an important role in the behaviour of the physical world. Somehow the very world of physical reality seems almost mysteriously to emerge out of the platonic world of mathematics.*"[8] What is clear is that information in Cyberspace is digital and non-material, in contrast to material space-time, which has analogue characteristics. Space-time exists in three spatial dimensions but information in Cyberspace would appear to have none. It can be argued that an algorithm or line of code is a one-dimensional string of information but it lacks any *spatial* qualities. Yet, both space-time and Cyberspace are logical, rational and, in quantum terms, 'local'. Cyberspace needs to operate in tandem with space-time to be able to impact our physical world as it is non-material, while computers act as an interface between these first two domains. Figure 7.2 illustrates some of the relationships between space-time and Cyberspace.

FIGURE 7.2 – MAPPING CYBERSPACE

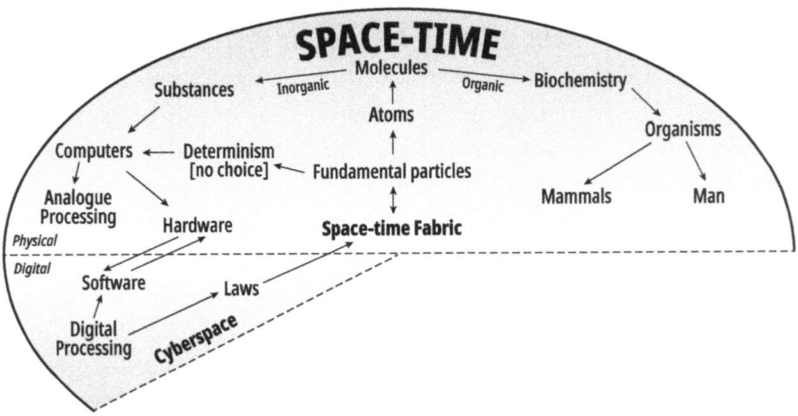

In addition, as Penrose suggests, the natural laws of the Universe appear to have similar characteristics to the deterministic processing we observe in computers in that they too are logical, rational & local. So, space-time and Cyberspace seem to operate together like conjoined twins, with Cyberspace acting as the non-material ('software') twin to its physical buddy, space-time. In marked contrast, the human mind works very differently to each of these.

7.4 – MINDSPACE

The third domain is Mindspace – the realm of thought, which differs significantly from Cyberspace. Research over the past century has shown that our memory operates holographically and is disseminated, indicating that Cyberspace and Mindspace are distinct.[9,10]

Modern materialistic science dictates that we can only do what our physical genes prescribe, which precludes free will, but we all experience the choices that life offers every day, so could it be that there is something more than the physical and that a purely material universe is not the complete story regarding our human existence. It seems that something is missing. Professor of Mathematics and Cosmology, Bernard Carr suggests that *"mental space"* exists in its own right and goes beyond physical space but that physical space is actually a subset of mental space.[11] However, the disseminated, complex meaningful information encountered in Mindspace seems different to space-time and has not been able to be described by overarching rules or formulae that might be generated by programmable algorithms, so has evaded credible mechanistic explanations. Instead, as discussed in Chapter 6, it would appear to require an intelligent mind to formulate the messages that originate there.

Further, in complex-brained animals, mind exhibits consciousness, permitting thought, while the human mind in addition exhibits self-awareness, which attributes meaning[12] that allows choice or free will.[13] In addition, as we saw in the previous chapter, the mind-brain complex is a compound entity that has both physical and quantum processing, unlike current computers which are deterministic and operate within Cyberspace alone. Through complex brains, Mindspace interfaces with space-time, yet it is discrete, as illustrated in Figure 7.3.

FIGURE 7.3 – MAPPING MINDSPACE

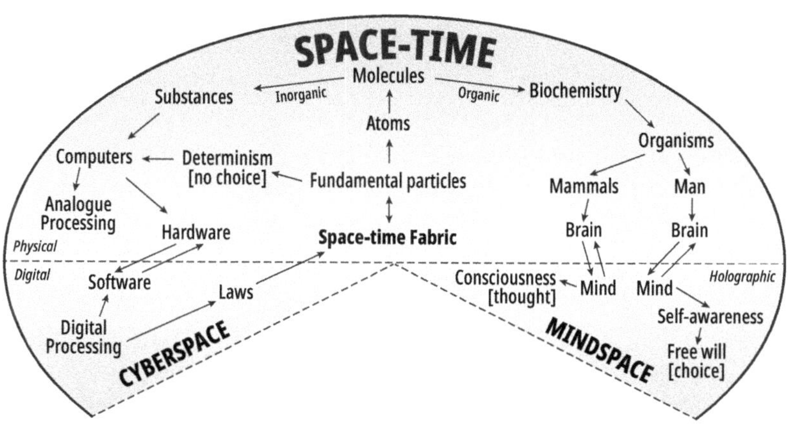

7.5 – THE QUANTUM DOMAIN

Last but certainly not least, is the fourth information type which is found in the quantum domain which defies logic and common sense but is once again disseminated and non-local. Despite a century of research, much of the quantum realm remains opaque to human investigation and hence is still poorly understood. However, the operation of modern electronics such as silicon chips, smartphones, global positioning systems and lasers are dependent upon it so cannot be ignored. This completes our information overview of the Universe (see Figure 7.4).

FIGURE 7.4 – INFORMATION MAP OF THE QUANTUM DOMAIN

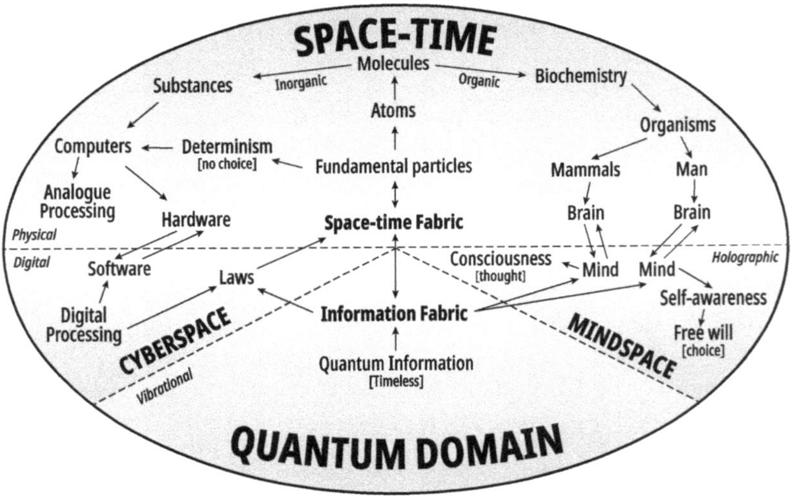

Underlying the space-time fabric there appears to be an *Information Fabric* which derives from quantum domain information that specifies and actualises the space-time fabric. However, the space-time fabric and the information fabric appear to be more like the opposite sides of a coin rather than distinct entities, where subatomic particles can exhibit both material characteristics through the space-time fabric and non-material phenomena from the information fabric simultaneously. The space-time/information fabric appears to be the interface between space-time and the quantum domain. The information fabric is also sometimes referred to as the quantum vacuum or the zero-point field. In addition, Figure 7.4 suggests that there are two primary platforms: space-time and the quantum domain, on which Cyberspace and Mindspace run as applications, respectively. Thus, each domain has differing attributes and possesses its own unique characteristics (see Table 7.1).

TABLE 7.1 – THE CHARACTERISTICS OF THE INFORMATION DOMAINS

	1	2	3	4
Realms	DETERMINISTIC		NON-DETERMINISTIC	
Domains	Space-time	Cyberspace	Mindspace	Quantum
Mechanics	Einsteinian Relativity	Shannon (syntactic) data management	Meaningful (semantic) info management	Quantum mechanics
Operations	Analogue (continuous)	Digital (discrete)	Holographic (dispersed)	Vibrational (probabilistic)
Character	Physical (particle/wave)	Granular (deterministic)	Disseminated (non-deterministic)	Fuzzy/blurry (smeared)
Function	Structural (platform)	Computational (processing)	Interpretive (free will & meaning)	Actualising (realizing)
Scope	Local		Non-local	
Formulation	Continuous	Discrete	Continuous	Quantised
Implications	Materially bounded	Non-material & timeless		Bounded by information

Time can be viewed as *a measure of change*, which requires discreteness. So, as the quantum domain is disseminated, all parts are permanently connected; therefore change and hence time are restricted to space-time.

Space-time is a Euclidean space which possesses finite dimensions. In contrast, quantum physicist, Sean Carroll, suggests that all of reality is part of a quantum waveform in Hilbert space which is an infinite-dimensional inner space that is complete or closed.[14] Figure 7.4, would also suggest that Mindspace might be within Hilbert space (although a different one to the quantum domain) as both are non-material and timeless.

It is interesting to note the similarities between Domains 1, 2 and 3 above and Karl Popper's Three Worlds Theory from 1968. His World 1, incorporating the physical equates to space-time. World 2 of thoughts and feelings corresponds to Mindspace and World 3, the products of mind, such as algorithms and theories, is broadly equivalent to Cyberspace.[8] Alistair McGrath points out that World 1 of the physical encapsulates mainstream science and to some scientists represents all that is 'real'. However, as McGrath notes, it is devoid of meaning. So, McGrath encourages us to include the metaphysical with the physical to encompass meaning.[15] Popper's theory was adapted and expanded in 1994 by mathematician and physicist, Sir Roger Penrose, who was awarded the Nobel prize in Physics in 2020. Penrose also identifies three separate realms: *"There is the physical world, (things like tables, ...) and there is the mental world, (that is our experience, consciousness, feelings about things ...) and there is the abstract world (and I would be more specific about that, it is a mathematical abstraction).*[16] Table 7.2 compares Popper's worlds with McGrath's analysis, Penrose's realms and the corresponding Information domains.

TABLE 7.2 – A COMPARISON OF POPPER'S THREE WORLDS WITH PENROSE AND MCGRATH

Popper	Penrose	McGrath[17]	Information Mechanics
World 1: Objective	Physical	Cognitive	Space-time
World 2: Subjective	Mental	Affective	Mindspace
World 3: Theoretical	Abstract	Affecto-cognitive	Cyberspace

Since the 19th century, reductionism has progressively teased apart and separated the various fields of academic study pertaining to our universe, dividing them into distinct disciplines. They can be thought of as stratified layers, as in a layered cake or a geological formation; these fields include: mathematics, physics, chemistry, cosmology, biology but also music, literature, philosophy, theology and the list goes on and on! This splintering has been further exacerbated by the current explosion of knowledge, so that each discipline has had to become more and more sub-specialized to cope with this accelerating accumulation of information. The result is ever narrowing fields of endeavour. Stratifying the information horizontally allows each discipline to spread out and understand the inner workings of their field of study but this often means that we lose sight of the bigger picture, whereas the human mind has the additional ability to re-align the information vertically (as in Table 7.2) in what Alistair McGrath is encouraging us to consider the re-integration of disciplines into a re-invigorated field of Natural

Philosophy. McGrath observes: *"Popper's three worlds can helpfully be distinguished, but they cannot be separated in the lives of individuals."*[17] Information mechanics similarly encompasses the horizontally stratified detail within each discipline but also accommodates the unity acquired through vertical integration that is possible by the amalgamation found when the multiple layers are recombined in modern academic study. In addition, it should be noted that Figure 7.4 indicates, as Sean Carroll has suggested, that each of Popper's three worlds is underpinned by the enigmatic stochastic quantum domain.[14] Following McGrath's line of thinking, despite the differences, Table 7.1 shows that there are some similarities too. Both space-time and Cyberspace are local and work together in the deterministic realm, while Mindspace and the quantum domain synchronise within the non-deterministic realm.

7.6 – DIMENSIONS

Having determined that there are different domains, we need to ask: 'where are they?' In *The Grand Design*, Hawking and Mlodinow propose eleven-dimensional M-Theory as the most likely structure for the Universe.[18] There are other theories but if Hawking and Mlodinow are correct, then M-Theory suggests a structure within which the domains described in Table 7.1 might be located. Einstein showed that space-time occupies 3+1 spatial dimensions, whereas a line of digital code in Cyberspace is a one-dimensional (non-spatial) string. Mindspace might occupy another two non-spatial dimensions (consciousness and self-awareness), leaving four to five dimensions in the quantum domain. In all these discussions, anti-matter has not been mentioned as there is not a lot of evidence from which to draw conclusions, except to say that it seems to be missing from our physical universe as when it meets normal matter, the two self-destruct. This suggests that the missing anti-matter may be in another dimension. This is purely speculative at the present but suggests directions for further study.

CHAPTER EIGHT

RECONCILING THE CLASSICAL AND QUANTUM REALMS

"...the supreme goal of all theory is to make the irreducible basic elements as simple and as few as possible without having to surrender the adequate representation of a single datum of experience." or *"Make things as simple as possible, but no simpler."*

– Albert Einstein (1933)[1]

In the previous chapter, the four different information domains were described according to their discernible characteristic *differences* but we have already established that our universe is also described by intricate mathematics, which operates seamlessly across the Cosmos. Figure 7.4 presents a relational map of our universe that links the four differing information domains but to reconcile these differing domains to understand how they work *together*, we need to dig deeper with the aim of integrating space-time with the quantum domain.

8.1 – MATHEMATICAL REPRESENTATION

Isaac Newton most famously demonstrated how mathematics can be used to model the behaviour of our universe in his *Principia Mathematica* (1687), however since Eugene Wigner wrote about: *"The unreasonable effectiveness of mathematics in the natural sciences"* in 1960,[2] more and more scientists have been describing mathematical relationships in the Universe. Indeed, since its creation in 2016, the L-functions and Modular Forms Data Base has recorded 5 *million* pages of mathematical formulae. In 2011, theoretical physicist, Leonard Susskind stated: *"...as you go deeper and deeper into reality, mathematics becomes the only way to describe reality..."* [3] while Max

Tegmark commented: "*…our universe isn't just described by math, I think it is math.*"[3] It is intriguing to note that √-1 cannot physically exist in material space-time and hence has been dubbed the "imaginary" or "impossible" number (denoted as 'i') but the quantum realm occupies the region where √-1 is *required* to make sense of underlying reality![4] Thus, √-1 splits the universal realms: in space-time it is 'impossible' while the quantum domain cannot exist without it. In mathematics √-1 occupies a region called the complex number plane. It is likely that other mathematical correlations will be made between the individual information domains over time. James Glattfelder sheds further light on this when he observes that: "*Quantum mechanics and general relativity are incompatible due to the different nature of their underlying 'programming' (i.e., the discrete vs. the continuous), exposing a missing feature...*"[5] Programming delineates the way information is utilised. So, the answer to the question of: *What is the meaning of life, the Universe and everything?* is not 42 but √-1 or 'i'. (Apologies to Douglas Adams who unfortunately was misinformed!)[6] Once again, this is an area that requires more research but now it is time to examine our next enigmatic feature of interest, the existence of black holes.

8.2 – BLACK HOLES

The subject of black holes goes back over 240 years, when they were first theorised in 1783 by John Mitchell (1724-1793), a clergyman in Yorkshire, England, before being described mathematically in 1916 by Karl Schwarzschild (1873-1916); yet, they are so strange that it was not until 1967 that they were given their current name by physicist, John Wheeler. They can form when a medium sized star comes to the end of its fuel supply and under the weight of its own internal gravity it implodes, shrinking very rapidly to a single point. The extreme gravitational pull of a black hole deforms space-time to the extent that it appears to puncture or tear a hole in the fabric of space-time itself, leading to the quantum domain. For the physicists, James Glattfelder explains that when the information content (entropy) of an object exceeds its holographic bound [too much information to store in the space available], a black hole forms and the excess information spills into the quantum domain. Another way of expressing this is that the information content within the rapidly shrinking star, which, being in space-time must be instantiated within a physical vehicle, exceeds the maximum information capacity allowable within the space-time fabric, which 'disintegrates' under the load, creating the black hole.[7] Black holes appear to be a one-way connection between space-time and the quantum domain. Hawking and Penrose noted: "*...a large amount of information is lost when a body collapses to form a black hole*"[8] but as information can never be destroyed, it would appear to exit space-time, entering the quantum domain.[9]

Black holes appear black because their gravitational pull is so great that not even light can escape and is sucked back inside it. Surrounding every black hole is an 'event horizon' at which point space-time appears to end. The black hole's event horizon is the portal between space-time and the quantum domain. Inside the event horizon, the equations of general relativity break down as the curvature of space-time becomes infinite (at least according to the mathematics).[10] In reality, no one has been there to objectively measure it but if they did they would not be able to come back to tell us due to the extreme gravity!

As an analogy, imagine space-time as Thor Heyerdahl's balsa-log raft, the Kon-Tiki,[11] that is sailing across the Pacific Ocean, which is analogous to the quantum domain. As he sails across the ocean, contemplate a huge storm that snapped off the mast, which speared through the deck and disappeared into the ocean below. There is now a hole in the deck but due to the inherent stability of the balsa logs, the raft stays afloat and carries on much as before, unless something on the deck gets too close to the hole and falls in, when it is lost forever in the ocean below. Note that it was the extreme conditions involved in the collapse of the mast that tore the hole in the fabric of the deck and so it is with black holes.

8.3 – WHITE HOLES

Gigantic black holes are now thought to lie at the centre of spiral galaxies too, such as our Milky Way, but no one is sure how they got there. However, any singularity can be viewed as a boundary between space-time and the quantum domain. Jim Al-Khalili, Professor of Theoretical Physics, writes: "*There are certain solutions of the equations of general relativity which allow for the existence not only of black holes but of objects called white holes too. But it turns out that this would be possible only if the universe had certain properties, called initial conditions, at the time of the Big Bang. These highly speculative objects are the opposite of black holes which, rather than sucking matter in, would spew matter and energy out into the universe. Their singularities would be different too and would mark the beginning rather the end of time.*"[12] So, the singularity involving the formation of the Universe at the 'Big Bang' may represent one (or many) white holes that ushered information and energy (light) into our nascent universe. However, Al-Khalili also points out that white holes are very unstable and convert into black holes in a matter of seconds after forming.[13] In 2022, the James Webb Space Telescope discovered unexpectedly large galaxies in the early Universe which the Standard Model predicts should have taken significantly longer to form. Could these be the remnants of white holes?[14] White holes would have had to have been specifically programmed to behave that way (for the introduction of primary programming information at the formation of the Universe to boot up the system), but as theoretical physicist, Neil Turok points out, for the Big Bang to work, without "*...the initial conditions and the potential energy to keep inflation going for a sufficiently long time...*" there is no realistic explanation for the formation of the Universe according to the Standard Model anyway.[15,16]

8.4 – UNIVERSE OR MULTIVERSE

The requirement for the existence of initial conditions for the formation of our universe has prompted the speculation that the birth of the Cosmos was auspiced by an even larger entity that spontaneously produces universes, dubbed the *Multiverse*. However, this merely pushes the initial conditions back one more step. If the Multiverse exists, all the information that specifies *our* universe must have derived from the pre-existing Multiverse which also needed specification. So where did the information come from that specifies the Multiverse? Some suggest that the Multiverse has existed forever and so never needed specification, however, that hypothesis has less evidence than the suggestion that: "*In the beginning God created the heavens and the Earth.*"[17] As the Many Worlds' Interpretation of quantum mechanics requires that a new universe is formed every time a choice is made or an alternative occurs within our universe (such as the spin on a particle changes from up to down), this multiverse suggestion fails the 'common sense' test, not to mention Ockham's razor. The trillions upon trillions of universes required is absurd and no one is even attempting to explain how this might realistically happen. In addition, the Multiverse needs a mechanism to periodically inject new information into *our* universe, such as to provide the DNA code for the first cell when life arose and to bridge the other information discontinuities. (See Chapters 4 & 9.) A mechanism for this within the Standard Model is beyond comprehension. The more rational conclusion is that a *single* universe was created by an *eternal* God, and this remains the inference to the best explanation.

8.5 – RECONCILING SPACE-TIME WITH THE QUANTUM DOMAIN

In 1924, French physicist Louis de Broglie (1892-1987) postulated that 'spherical standing waves' could explain the wave effects in particles which is also referred to as a 'matter wave'. This is the basis of the wave-particle duality of matter for which he was awarded the Nobel Prize in Physics in 1929. In 1927 de Broglie developed his ideas into a causal interpretation of quantum mechanics which he called the Pilot-wave Theory. Unfortunately, it was not well received and the Copenhagen interpretation of quantum mechanics championed by Niels Bohr (1885-1962) prevailed until 1952, when David Bohm (1917-1992), at the suggestion of Albert Einstein, extended de Broglie's theory in what is now the de Broglie-Bohm Theory of Quantum Mechanics which is deterministic and avoids the Copenhagen requirement for an observer. It has been highly controversial but has been growing in favour over the past few decades.

However, despite the controversies over what the mathematics means in reality, and the apparent fundamental differences between space-time and the quantum domain interpretations, in the real world, they co-ordinate seamlessly. This implies that it is our *understanding* of the theories that is not yet fit for purpose. The No-Hiding Theorem from Chapter 4 suggests that the foundational information elements of matter exist in the quantum domain in a state of entanglement that includes all potential outcomes. However, when these enter or interact with space-time, they lose some or all of their extra information (perhaps even extra dimensionality) that is expressed in the quantum domain but collapse into a material or analogue state in space-time, with fewer possibilities and henceforth behave deterministically while in the space-time domain. When we measure fundamental particles, we cause them to interact with the space-time domain through our measuring device. It would seem that it is not the act of measurement or observation but the forced interaction with space-time (or more specifically, the space-time fabric) that collapses the quantum state through the constraints or limitations of the laws of our physical universe, within our four-dimensional material domain. Without the extra information/dimensions necessary for their full expression (as in the quantum domain), the particles being studied can only express a reduced set of behaviours, as seen when the information in a matter-wave collapses to a particle in the double-slit experiment. As it interacts with the space-time domain, the information in its quantum dimensions is constrained or incapable of its full expression and the waveform collapses into an analogue form, a process termed: 'quantum leakage'.[18]

8.6 – QUANTUM LEAKAGE

Quantum leakage also seems to occur when the mass of an entity exceeds a certain size. In 1999, a team of Austrian physicists fired a cluster of sixty carbon atoms, called 'fullerenes' (or 'buckyballs'), through a double slit and found that the buckyballs recreated the same classical interference patterns as light waves do.[19] Buckyballs therefore seem to be small enough to retain their quantum domain properties but larger entities, such as tennis balls (or Schrödinger's cat) do not; larger objects cross a threshold, above which the quantum superposition can no longer be sustained and once again the quantum wave-form collapses to its material state. Jim Al-Khalili writes: "*Quantum mechanics dictates that the many trillions of atoms that make up* [a macroscopic object such as] *the measuring device … must also exist in a superposition. However, these delicate quantum effects become too complex to be maintained in such a large macroscopic device and leak away, almost like heat dissipating from a hot body. The process is known as 'decoherence' and is currently a subject of much debate and research. ... It takes place whenever a quantum system is no longer isolated from its surrounding environment (which can be anything from a Geiger counter to a lump of rock or even surrounding air molecules;) ... If the connection to its external environment is strong enough, then the original delicate superposition gets lost very quickly. In fact, decoherence is one of the fastest and most efficient processes in the whole of physics.*"[20] Professor Gavin Brennan, Director of the Macquarie University Centre for Quantum Engineering in Sydney explains: "*The more things you have to control, the more things they can interact with. And the things they interact with get information. Your quantum system leaks information and when it does that it loses the properties that make it quantum.*"[21] The precise mechanism for decoherence is yet to be discerned but there appears to be a leakage of information from quantum systems when they come into contact with the space-time domain; without this extra information, the quantum system being studied loses its quantum characteristics. Planck length appears to determine the zone within which decoherence occurs, as it is the point at which space-time begins. Entities less than Planck length seem to 'pass through' the pixelations of space-time but those that are larger than 10^{-35} metres become 'enmeshed' in the pixilation grid which triggers the waveform collapse. This size specification seems to be an intrinsic property of quantum domain information.

The information in the quantum domain appears to be more richly configured when it resides there as it exists in a state of superposition. Some of that information seems to be 'lost' (which is to say that it either disperses into space-time or retreats into the quantum domain) when we measure it in space-time. This limits the information able to be displayed and hence interpreted in space-time, which means that we have to settle for a *probability* of the actual value of some of the information in the quantum domain unless it could perhaps be decerned from Mindspace (which can be seen in Figure 8.1 to be a sub-set of the quantum domain). It is intriguing that the human mind-brain complex appears to be able to cross the divide between these two realms. The human brain is a material entity so must operate within space-time, which is evidenced in the classical firing of the neurons within the brain (as seen in the recording of the electrical activity of the brain in an electroencephalogram or EEG) but if the intracellular function of nerve cells operate within the quantum domain, as Roger Penrose and Stuart Hameroff suggest (as discussed in Chapter 6), then the human mind-brain complex may operate on two levels simultaneously, where the macular-functions of neurons operate in material space-time while the intracellular nano-functions of the cytoskeleton function within the quantum domain. How these two levels of macular and nano work together is yet to be fully determined but this model offers a possible mechanism for the observed operation of the human mind, such as the motor functions being distinctly regional, plus the well described left and right brain phenomena detailed in all neurology textbooks and yet memory is demonstrably holographic as established by Karl Lashley and Karl Pribram (see Chapter 6).

However, due to quantum leakage, not every entity that can be conceived of in Mindspace is likely to be able to be represented in space-time due to the different functionality of space-time and the quantum domain, a case of 'mind over matter'! This might lead to a prediction to test the theory. Alistair McGrath points out that facts in space-time are very different to theories, which are in Mindspace; to conflate them is to commit a category error.[22]

8.7 – EMOTIONS

A further intriguing question, following this line of inquiry, is that, as mentioned in Chapter 6, most cells contain cytoskeletons and it was previously pointed out that this may be the way that simple organisms like amoebae can process information that directs them towards food or light. If amoebae can process information in a single cell, this begs the question as to whether all the rest of mammalian cells are not processing quantum information in the same way? Non-brain cells understandably do not seem to process thoughts as they lack sufficient bundling of their microtubules, let alone a physical brain or neurons, the hardware that seems necessary in space-time to think consciously, however that does not mean that they do not process other information that might be valuable. Emotions have traditionally been considered to be the result of a functioning brain, however could it be that the rest of the cells in our bodies might not contribute to the generation of emotions? In English we have a number of terms such as: "gut feelings", "I feel it in my bones" and our emotions colloquially are often attributed to our *heart* rather than our brain. At times of extreme emotion, we can experience a frisson or 'thrill' which is like a transient tingling all over our body which might be mediated in this way. There certainly are mechanisms that support the idea that our emotions are *interpreted* via our brain, being auspiced by the neuro-endocrine system, however a precise mechanism explaining their *source* has not yet been explained. Might it instead be that thoughts originate in the mind-brain complex, while emotions emanate from somatic cells, yet be mediated via the neuro-endocrine system, part of which involves a relay back to our MBC? At present, sufficient information is lacking.

8.8 – OUR HOLOGRAPHIC UNIVERSE

In an article in New Scientist in 2024, Thomas Lewton writes: "*Netta Engelhardt of the Massachusetts Institute of Technology... and her colleagues are deciphering the enigmatic connections between the quantum realm and classical gravity - and reinforcing the revolutionary idea that the latter emerges from the former like a holographic projection. ... The idea is that higher-dimensional space-time emerges from a lower-dimensional realm, described by quantum theory, in a way roughly akin to how a hologram is projected from a flat surface. You can translate between these two equivalent descriptions of reality using what you can imagine as a classical-quantum dictionary... Space-time is thought of as emerging from a flat surface, governed by quantum mechanics, as if it were a hologram. ...* [The difficulty is to] *grasp the convoluted way in which the holographic dictionary translates the quantum world into gravity. That is difficult because, instead of a one-to-one translation, each patch of the bulk universe seems to be encoded on the holographic plate,... more than once, in different places."*[23]

Hungarian scientist, Denis Gabor (1900-1979) developed the concept of holograms in 1947 for which he was awarded the Nobel Prize in Physics in 1971. However, the first holograms were not created until 1962, shortly after the development of lasers. A true hologram is like a three-dimensional photograph but with a difference. It is an image where the information that describes the object is distributed or disseminated over a two-dimensional medium that stores it. The recording medium that most people are familiar with is a sheet of photographic film. A hologram is created by shining a laser beam at an object and recording its interference pattern. Although the image is recorded on a two-dimensional surface, when illuminated by the same wavelength of laser light that created it, it generates a three-dimensional image. A hologram appears to have three dimensions when viewed in space-time but it is in fact complex, mathematically-encoded information that is recorded in only two dimensions, which is why it can be stored on a flat sheet of film. The *Holographic Theory of the Universe* proposes that the entire Universe is similarly encoded on an abstract, timeless two-dimensional surface. In trying to visualise this, imagine the 2-dimensional curved surface of a piece of paper that has information recorded only on one side, that has then been curled into a cylinder or a sphere to enclose a three-dimensional space.[24] Now, think of space-time as a hologram but instead of being constructed from projected light, think of an analogue projection of matter – a 'mattergram'! This makes space-time physical, while its foundation is still the result of the projection of information via mathematical algorithms or rules but from the quantum domain.[25]

8.9 – NEEDING TO DIG DEEPER

We have covered a lot of material in a couple of chapters which may have left you feeling slightly overwhelmed. Jim Al-Khalili, summarised some of these ideas when he wrote: "*It is just possible that quantum mechanics is not the whole story, and that the unpredictability of a process such as radioactive decay is indeed due to our ignorance. What we may be lacking is a deeper understanding of nature whereby we would, after all, be able to predict exactly when any given atom might decay, in principle if not in practise, … If this is the case we may need to go beyond quantum mechanics to find the answer, or at least develop a different interpretation of the quantum rules. Einstein himself held this view, and this is what he meant by his famous remark that 'God does not play dice'. Einstein could not quite accept the randomness of the quantum world.*

Although Einstein's version of the argument has since been shown to be wrong, there is another way of interpreting quantum theory that does not conflict with the standard version, but according to which the subatomic world behaves completely deterministically. It originates with the work of a physicist by the name of David Bohm and has been known for over half a century.… According to Bohm, the quantum world's unpredictability arises not from true randomness but from the existence of information that will always remain hidden from us and without which we cannot make precise predictions. The quantum world is unpredictable not because we can never burrow down deep enough into it, … but rather because we are simply unable to probe the subatomic world without disturbing it in some way.… In David Bohm's version of quantum theory, every particle in the Universe has a quantum force field that controls its actions; by measuring the particle, we disrupt this field and so change the particle's behaviour.[26,27]

This leads to another perplexing question posed by New Scientist: "*Is reality made of particles or fields? We talk of particles, which is to say things that occupy points (more or less) in space - yet particles are just energetic spikes, or "excitations", in underlying quantum fields. A photon, say, as the carrier of electromagnetism, is an excitation of the underlying electromagnetic field. These fields permeate the entire universe, which means that there is some residual photon-ness everywhere - so-called virtual particles, which appear fleetingly and then vanish - even when there isn't enough energy around to manifest a particle proper. Fields, then are usually considered more fundamental. The math simply doesn't work without them. But the ambivalent quantum nature of a field means that it can't be observed directly: any observation requires a dose of energy that, in effect, forces the field to show its hand and produce a particle. For this reason, particle talk seems unavoidable. Particles are what we see."*[28] So, the answer to the original question is that quantum reality involves fields which translates into space-time reality as particles. This understanding offers us a model for tunnelling; the particle in space-time appears to jump through impenetrable barriers but in reality, it is the quantum field behaviour in the opaque quantum realm where the 'sleight of hand' is actually happening.

So, in summary, the picture is emerging where space-time appears to be an emergent property (a derivative or subset) of the larger underlying quantum domain. In turn, the quantum domain is constructed from an information fabric (that could just as easily be described as an information field), which would seem to have been infused into the Universe following its formation (as discussed previously). Further, if the Holographic Theory of the Universe proves to be true, this would make the quantum domain the default realm for the functioning of our universe and space-time as the derivative entity. As such, the more limited functionality of space-time is not surprising as it is secondary. We therefore need to reverse classical notions of what the 'normal' state of an atom is (or any entity for that matter) and view the superposition of atoms in the quantum domain as the primary or principal form. It would seem logical therefore, that left to their own devices, an unperturbed atom will exist in or revert to its quantum configuration unless constrained by an association with space-time. An analogy is the operation of mobile phones, where a 5G (5th Generation) phone defaults to 5G operating mode when within a 5G network area but will operate, albeit with reduced functionality, if it finds itself in a 4G or 3G network. However, once it enters a 5G network once more, it instantly reverts to the higher 5G functionality.

8.10 – PROBABILITIES AND THE QUANTUM REALM

One of the most perplexing characteristics of the quantum realm is the necessity to use probabilities to work with objects within it. This requirement left Einstein so perturbed that he could never fully embrace it even though he was partly responsible for describing it. The biggest problem is that nearly all physicists agree that the mathematics works but there is scant agreement on what it means. In Chapter 7, it was seen that space-time emerges from the space-time fabric which in turn appears to be an interface between the quantum domain and space-time itself. It would be convenient if the space-time fabric was part of Cyberspace, but these ideas are not contingent upon it and there may be other spaces not yet identified too. What is more significant however, is that Batiz and Chauhan in their 2006 paper: *Holographic Principle and Quantum Physics* conclude that the space in space-time is quantised, meaning that it can be thought of as being like a three-dimensional pixilated digital television screen. Further, they conclude that the resolution of these 'pixels' is: "*... plank scale, which can be perceived as the minimal distance between two points in the quantized space-time continuum.*"[29] This explains why the minimum resolution of matter in space-time is plank length, which is 10^{-35} metres. Below this size, matter has no resolution as it slips back into the quantum domain (via the space-time fabric). Think of pixelated space-time as a sieve and ultra-microscopic entities simply fall through the 'spaces' between the pixilated grid lines, which lands them in the quantum realm.

However, the mathematical descriptions of space-time and the quantum domain have very different viewpoints. Professor of Cognitive Psychology, Donald Hoffman writes: "*General relativity demands that space-time, the four-dimensional structure that space and time form together, is smooth and continuous, whereas a quantum description requires it to be pixelated.*"[30] This apparent contradiction can be understood by realising that general relativity is 'in' space-time, looking down, and from a macroscopic perspective, space-time does indeed seem smooth and continuous due to the enormous difference between the macro and the nano scales, but quantum mechanics is 'in' the quantum realm, looking up and from the ultra-microscopic perspective, sees space-time as a pixelated space. An analogy is that when we are on Earth, Mount Everest seems very high but if we are in a space capsule orbiting the Earth, the surface below us appears smooth. This is a frame of reference issue. Remember that we were previously told that the programming of the two realms is different. So, a functional or structural map of these domains can now be constructed as shown in Figure 8.1.

FIGURE 8.1: A STRUCTURAL MAP OF THE COSMIC DOMAINS

This is a two-dimensional diagrammatic model of our four-dimensional universe (and maybe eleven), that scales over many levels from Planck scale to our macular experience of the Universe so the scales are not comparable but it illustrates the relationships. Also note that this diagram portrays a static picture of the Cosmos but it is important to remember that all the domains are in a constant state of flux. The bold dashed line between Cyberspace and the Quantum Domain represents 10^{-35} metres which is the distance light travels in one unit of Planck time (10^{-43} seconds), so the speed of light also appears to be tied to the quantum information that specifies space-time.

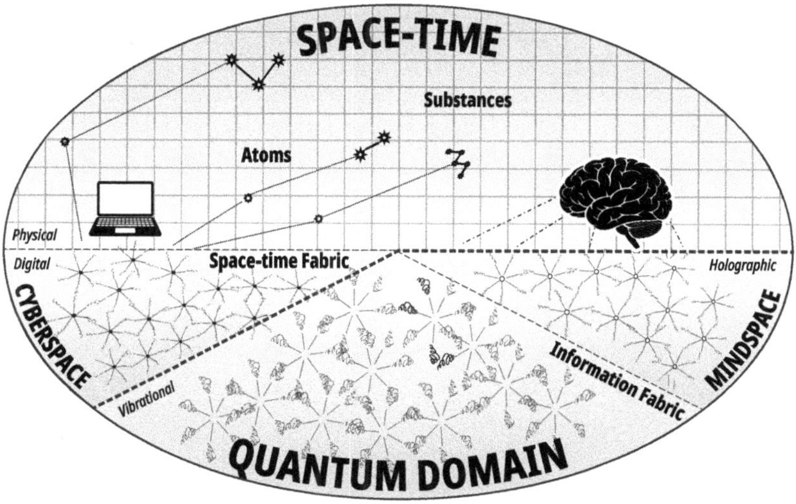

When thinking about the Quantum Domain, theoretical physicist, Matt Strassler suggests that we can be misled by the nomenclature we use, such as describing electrons and other fundamental entities as 'particles' and/or 'waves', because of their everyday use in space-time. He suggests that when we consider *quantum scale behaviours*, the term *"wavicles"* can help us to consider their true nature as quantum entities. This helps us to distinguish in our minds between material space-time behaviours and the language necessary to understand quantum-thought and quantum-speak. So, think of the quantum domain as being an interconnected ocean of wavicles![31]

Another quantum event with which we struggle is the randomness of radioactive decay, which is a quantum event that has not yet been explained. To understand the problem, think of the quantum domain as an ocean, where all the water molecules are interconnected and constantly moving, while a photon is like a wave that might appear from time to time on the surface of the ocean. It will come and then disappear again, driven by unseen energies, like the ocean currents, that are beneath the surface of the ocean. Schrödinger's equation very accurately describes what is seen on the surface but what is missing are the information rules or algorithms that determine the hidden activity under the surface. Until these opaque information rules are understood, as David Bohm suggested, the quantum domain cannot be properly understood.

So, to summarise, space-time is analogous to a 3-dimensional pixelated television 'screen' that is driven by de Broglie matter-waves, which broadcast a three-dimensional holographic "mattergram'. This process is driven by the space-time fabric which is equivalent to the 'transmission station' while the quantum domain is the 'television studio' where the content is being created and stored. This explains the No-hiding Theorem, as any information that is lost in space-time still has the original copy in the quantum domain (if we only knew how to retrieve it)!

8.11 – FINDING THE ELECTRON

In marked contrast to space-time, which has a pixelated 3-D materialistic grid construction, the quantum domain is a holographic disseminated, interconnected whole, where nothing is discrete, which explains Einstein's: *"spooky action at a distance"*. The electron is somewhere in the whole but it is impossible to say exactly where until it pops up and reveals itself in space-time. Its distributed nature can be thought of as a woven mat (the information fabric) in the centre of a polished wooden floor (the quantum domain). On the woven mat is a smaller plastic shower mat (the space-time fabric) with 'drainage' holes, through which a small pebble can fall onto the woven mat below. Most of the underlying mat is obscured but with a pair of tweezers, a thread can be pulled up from the woven mat through the shower mat holes. Now imagine that thread is a de Broglie standing wave which can only be successfully "grasped" by the tweezers if it is in the exact middle of the standing wave/woven fibre. David Bohm describes this as the 'wave-centre' of the disseminated matter wave. The likelihood of the success of this process could be calculated if enough detail could be known about the system but groping through the small hole in the shower mat and not knowing the direction of the woven thread below would, at best, allow the calculation of the probability that at any one time the middle of the thread/standing wave would be precisely grasped to capture it. However, to make matters more complicated, in actuality, the whole system is dynamic so rather than the mat lying in the middle of a wooden floor, imagine it is floating in a swimming pool (the quantum domain) with a high-capacity pump creating continual turbulence. Now, if the fibre is replaced by an electron, it can be appreciated why only probabilities can be calculated for locating disseminated entities within the quantum domain while viewing from space-time. The answer may be different if we could view these events from the quantum domain but currently there is no known way to do this.

Another way to imagine the superposition of particles is to appreciate that although they seem to be in two places at once, we are actually just seeing two parts of the same wavicle that is smeared out but still connected via the information fabric. Decoherence happens when the quantum link is broken. Einstein pointed out that information cannot travel faster than light, however this only applies to space-time due to its pixelated material characteristics, but the quantum domain is permanently connected so information transfer is instantaneous in the quantum realm.

And then there is the question of whether gravity is a quantum phenomenon. Nick Huggett, a philosopher of physics and Carlo Rovelli, a theoretical physicist are investigating the problem of quantum gravity or more precisely, *"gravity's quantum nature"*. They write: *"We already have plenty of observations about gravity's effects on the quantum behaviour of matter. Albert Einstein's theory works fine in these situations, from stellar dynamics, to the cosmological formation of galaxy clusters, all the way to lab experiments on the effect of Earth's gravity on quantum systems. But in all these scenarios, gravity itself behaves in a way that is consistent with classical physics; its quantum features are irrelevant."*[32] So, in the light of space-time being a 'matter-gram' projection from the quantum domain and space-time being in the classical realm, could it be that although quantum particles may be influenced by the gravity in space-time, the classical programming in space-time means that gravity itself is not a quantum phenomenon?

This would seem logical since gravity seems to emerge from the space-time fabric rather than the quantum domain directly. Recall that the 'programming' of the two realms is very different. (Also see Section 9.9 where the mathematics suggests that gravity is emergent.)

Mindspace is a subset of the quantum domain but has some differing features, so its structure is likely different again; the differing behaviour of the four domains suggests that their fine structure differs also. Memory appears to be one of the ways that Mindspace differs from the other three domains. By changing the angle or wavelength of a laser, multiple holograms can be stored in a single photographic sheet, which suggests the possibility that memory is stored through the holographic functioning of our quantum mind-brain complex. It is relevant to note that holograms can be created by different types of waves, not just by light.[33] In thinking about memory, remember that there is both short-term and long-term memory and it may be that short-term memory is chemically (physically) mediated in space-time, as it attenuates over minutes, days or months but longer-term memory is very likely holographic (which means it is mediated in the quantum domain) as it can survive decades and not fade.

8.12 – PLASMA

Another significant discovery of David Bohm's early research was regarding plasmas, a fourth state of matter after solids, liquids and gasses, where the electrons in superheated gasses stop behaving as individual entities but instead behave as if they are part of a larger interconnected whole, a single entity.[34] A possible explanation for this is that above a certain temperature (energy density), space-time and the space-time matrix would appear to breakdown or dissolve so that the plasma effectively behaves as if it were in the quantum domain once more. To continue the analogy of the Kon-Tiki, it is as if the fire for cooking the meals on the raft became too hot and burnt up the balsa logs so that the raft was completely destroyed leaving the contents (metal stove, cooking pots and utensils etc) immersed in the ocean.

Despite this explanation of the composition of our universe being more complex than previous ones, McGrath cautions us to be careful not to reduce the complexity of nature to the size of our minds but rather, to enlarge our minds to fully comprehend the richness of our universe.[35]

This information classification system provides explanatory power that gives insight into quantum leakage as well as the functioning of mind incorporating consciousness, thought, freewill, sleep and perhaps even emotions and unifies the seemingly disparate deterministic space-time and non-deterministic quantum domains into a functional universe that can support sentient life as we observe it. This modelling is still very incomplete and this is only the beginning of a new direction in science and philosophy but it has the potential to bear much fruit.

CHAPTER NINE

CONSTRUCTING THE COSMOS
FROM INFORMATION

"...a final theory must be concerned not with fields, not even with spacetime, but rather with information..."

– Jacob D. Bekenstein (1947-2015)[1]
Professor of Theoretical Physics

In Chapter 7, the four information domains were linked together to create a relational map of the Universe, while in Chapter 8, a mechanistic map was developed to form a functional plan of how the domains might work together. In this chapter, all the data points that have been previously identified are integrated together to produce a unified information map of the Cosmos, which was previously identified as the Infosphere.

9.1 – A SUMMARY OF OUR CURRENT PARADIGM

At the formation of the Universe, the Standard Model of Cosmology says that there was initially nothing until, in an abrupt (assumed) explosion, the Universe burst into existence without explanation. This presumably included the space-time fabric and all the precisely calibrated laws and constants that we can measure or calculate today that permit the Universe to function so beautifully. This would have required a vast amount of information which is never acknowledged or accounted for and so the 'Big Bang' defies logic and hence science. So, the formation of the Universe cannot be adequately explained by science, thus it is a *philosophical* theory. For a scientific explanation, at the very least, the appearance of the information must be accounted for.

The same requirement for the prior existence of information applies to the appearance of life and indeed, for every increasingly complex genus in Darwin's Tree of Life, such as bacteria, fish, birds, mammals etc. A robust explanation must include a method to explain the generation of this meaningful information that is everywhere in the biosphere and across the entire Universe.

This information is codified within the Universe in different ways. We usually think of a code as a series of letters or numbers but in life we find it in a DNA chemical string, while in the Cosmos, we identify it in the laws of the Universe, such as gravity. In the space-time domain we might find it in rocks as a chemical lattice; in Mindspace it might be a train of thought, whereas in Cyberspace it might be a formula or an algorithm. The quantum domain is rather opaque and difficult to see into, but we identify its activities by its outcomes, such as entanglement and tunnelling. Our current theories struggle to account for this information, which leaves huge gaps in our science that we need to explore further.

9.2 – INFORMATION GAPS AND DISCONTINUITIES

It would be logical to expect that the natural laws of the Universe were laid down at the inception of the Cosmos, but it does not appear that *all* the information required to specify life as we know it existed at the commencement. Michael Denton observes: "*Nature is stubbornly discontinuous, resistant to all attempts to reduce her to a Darwinian functional continuum.*"[2] Niles Eldridge and Stephen Jay Gould described these discontinuities in 1972. This is an information problem. Instead of the information being delivered in a steady stream throughout the history of life on Earth, it seems to have appeared episodically. This suggests that the information came in sequential software downloads and/or upgrades along the way, to specify the changes from non-life to life and from consciousness in vertebrates to self-awareness in human beings.[3] This has left gigantic gaps in our scientific knowledge. Table 9.1 summarises these steps or discontinuities.

TABLE 9.1 – UNSOLVED INFORMATION MYSTERIES OF THE UNIVERSE

	Process	Minimum information requiring specification
1	Formation of the Universe	Natural laws of the Universe and the space-time fabric
2	Appearance of energy/matter	Light/Sub-atomic and atomic particles
3	Appearance of a minimum set of organic molecules	RNA, DNA, proteins, fats and carbohydrates
4	Formation of functioning cell units	Cell membrane, control systems, nano-machines, cytoplasmic organelles, transport systems
5	Ability to reproduce	Transferable information code (such as the genome)
6	Arrival of single-celled life	Independent self-regulation (integration of all the above)
7	Arrival of multi-celled life	Complex information code (including an epigenome)
8	Arrival of complex life	Organ specialisation: nervous & digestive systems
9	Arrival of vertebrate life	Brain, consciousness and thought
10	Arrival of human beings	Self-awareness

Each of these steps requires the development of new information protocols, many of which are complex and beyond our ability to reproduce in a laboratory today, let alone conceive of having occurred naturally by 'chance'. Note that each level is a rate-limiting step. So, to sweep it all under the carpet and assert that without a credible information audit trail, the 'Big Bang' followed by Darwinian Evolution has achieved this monumental feat is anti-scientific. It also defies the logic required for philosophy. A new paradigm is needed to explain the 'miraculous' appearance of increasingly complex information systems to plug these information gaps between each stage as the Universe progressed from step 1 to step 10.

These gaps are due to *lack of* information. The crucial question is: whether the information exists but is yet to be discovered (gaps of ignorance) or the information never existed, except in the model of a theory that is being tested. So, is the gap revealed by our acquired knowledge of science or simply our ignorance of the Universe that is yet to be uncovered? As meticulously researching the Cambrian Explosion for the past century has shown no credible precursor fossils to support this event having occurred via Darwinian evolution, it is our *knowledge of science* that demonstrates the information gap and it is logical to propose that the relatively sudden appearance of the information specifying the organisms in the Cambrian Explosion was more likely to be the result of a different mechanism, such as an information download. If that was the case, then continuing to look for the organisms which preceded the new creatures that appeared at the beginning of the Cambrian period (such as Trilobites) will be fruitless. No matter how long the search continues, they will not be found because they *never* existed. This is a *structural gap* of science. However, taking a different example: before 1900 there was a complete lack of understanding regarding quantisation, so the problem of photons of light being discrete entities was a mystery until advances in our knowledge of science ultimately closed this *gap of ignorance* in our understanding. This is *science of the gaps* and as John Lennox points out: *"science … thrives in the gaps"*.[4] Indeed, the next Nobel Prize is waiting to be found in a gap in our scientific knowledge.

9.3 – FRONT-LOADING THE UNIVERSE

The Standard Model of Physics describes the Universe as beginning at a singularity when it was formed, and light (energy) entered our universe. We could say that the Cosmos was 'front-loaded' with light. However, there must also have been information, as the laws of the Universe appear to date from the inception of the Universe too and it is widely acknowledged that these laws are finely tuned for life.[5,6,7] Paul Davies also points out that these laws are information statements.[8] However, if there was one singularity, John Lennox asks, could there have been more, to account for the serial downloads?[9] At the very least, one could postulate, the appearance of plant life, the arrival of animals, the beginning of consciousness and the coming of self-awareness in human beings. The downloads would need to be sequential because the information code in life requires RNA and/or DNA which was not present when the Universe was formed. Significant preparatory work needed to take place before the material platforms (such as our Earth with its propitious biosphere) would be available to host life as we know it. These concepts of downloading information into our nascent cosmos may sound like *'Brave New World'* thinking but Lennox points out that Darwin's co-sponsor of the Theory of Evolution: *"Alfred Russell Wallace also thought that there were three … discontinuities: the origin of life; the origin of consciousness; and the origin of specifically human faculties."*[10]

9.4 – EVOLUTION OF THE GAPS

At this stage, I am sure that there will be one or two readers who are thinking that invoking extra singularities sounds a lot like the *"God of the gaps"* argument cited by Paul Davies (among others), where the hand of God is invoked for every scientific unknown.[11] To assess this, for each information gap we need to evaluate if it is a structural gap or a gap of ignorance. Does the Neo-Darwinian Synthesis or some other explanation such as an information paradigm give a better explanation? Unfortunately, due to the lack of evidence to derive detailed explanations, Neo-Darwinian theory still has a lot of issues that are too often plugged by 'just so' stories. Nobel laurate, Robert Laughlin commented: *"Much of present-day biological knowledge is ideological. A key symptom of ideological thinking is the explanation that has no implications and cannot be tested. I call such logical dead ends anti-theories because... they stop thinking rather than stimulate it. Evolution by natural selection, for instance, which Darwin conceived as a great theory has lately come to function as an anti-theory called upon to cover up embarrassing experimental shortcomings and legitimise findings that are at best questionable and at worst not even wrong. Your protein defies the laws of mass action - evolution did it! Your complicated mess of chemical reactions turns into a chicken - evolution! The human brain works on logical principles no computer can emulate? Evolution is the cause!"*[12] Lennox has dubbed this: "*...evolution of the gaps...*"[13] Fred Hoyle found 'evolution of the gaps' so unconvincing that he proposed 'aliens of the gaps', although, he called it: *'panspermia'* and this was without a shred of evidence.[14] Even so, the Search for Extra-Terrestrial Intelligence (SETI) has spent millions of dollars over the past sixty years, diligently searching but (to date) without finding a single piece of evidence. Evolution of the gaps can also be thought of as 'naturalism of the gaps'. The most obvious example is a Darwinian explanation for the formation of the first cell. Richard Dawkins writes that: "*...We have no evidence bearing upon the momentous event that was the start of evolution on this planet. It could have been an event of supreme rarity. It only had to happen once, and as far as we know it did happen only once.*"[15] The assumption is that: 'as we are here, it must have happened!' 'Junk' DNA is another example of evolution of the gaps, where the large amount of DNA in the human genome that did not code for proteins was touted for decades as 'evidence' of random 'mistakes' left by Darwinian evolution. When the ENCODE

Project showed that the vast majority of this is the management system, now called the 'epigenome', the Junk DNA hypothesis has been quietly shelved. The Human Genome Project dispelled another myth about the spontaneous ordering of living systems too. In 1998, Paul Davies wrote: *"The molecular traffic within the cell is essentially chaotic, driven by chemical attraction and repulsion and continually agitated by thermal energy. Yet out of this blind chaos order emerges spontaneously."*[16] Only a few years later, however, it was discovered that it is not 'blind chaos' that orders the cell but the epigenome (regulome) which comprises up to 96% of our human DNA and is now seen as the operations manual for life. This research is ongoing and much debated but blind chaos is no longer considered a significant factor in cell function.[17]

9.5 – MISNOMERS

The misnamed 'backwards wiring of the human eye' is yet another example where, what was initially thought to be an inefficient architecture of the eye that would negate a design hypothesis, has been found, with further research published in 2015, to be the most effective optimisation of colour vision by day, in addition to low-light vision at night.[18] Further, the human eye is optimally sensitive, being able to recognise even a single photon of light.[19] There is a long list of other examples of erroneous assumptions, including the misattribution of the purpose of the human appendix, which provides a repository for essential bacteria in the bowel to facilitate faster recovery after severe gastrointestinal infections such as cholera.[20]

Long-term *macro-evolution,* as described by Darwin to explain the expansion of his Tree of Life, requires both adaptation *with* increased complexity in order to accrete new information. The changes seen in bacteria that develop resistance to antibiotics is certainly adaptation but is the result of the random breaking of intra-cellular machinery that leads to the *loss of function* and *decreased information/complexity,* which is sometimes called *micro-evolution.* This can better equip that organism to survive in its niche in the biosphere where it must battle antibiotics, but it will not build a more sophisticated organism.[21,22]

In an interview in 2009, Richard Dawkins identified four entities with which evolution still struggles: the origin of life, the origin of sex, the origin of consciousness and the rise of morality.[23] Could it be that after

150 years of diligent research to close these gargantuan gaps, that they are examples of the introduction of information rather than a lack of understanding of Darwinian mechanisms?

9.6 – PUNCTUATED EQUILIBRIUM

As noted above, Niles Eldridge and Stephen Jay Gould described the Theory of Punctuated Equilibrium (evolution) in 1972 to explain the observed discontinuities in the fossil record. These obviously required the input or injection of new information, most notably at the Cambrian Explosion.[24] For decades Eldridge and Gould tried to demonstrate that a form of 'rapid evolution' could deliver a naturalistic solution but failed. In 2002, Gould wrote: "*... the tale itself illustrates the central fact of the fossil record so well - geologically abrupt origin and subsequent extended stasis of most species ... anatomy may fluctuate through time, but the last remnants of a species looked pretty much like the first representatives.*"[25] It now appears that their observations were correct but that their sought-after Darwinian mechanism was inadequate.[26] They needed a download of new information; so after millions of scientists have conducted intensive research for more than a century and found no credible Darwinian explanation to account for these discontinuities, it would not seem inappropriate to conclude that the inference to the best explanation is that the extra information in the body plans of the 'new' fossils was downloaded from an entity outside of space-time as these appear to be structural gaps rather than gaps of ignorance. However, as we have previously seen that meaningful information needs to be actively generated, we need to identify the information generators.

9.7 – INFORMATION GENERATION AND ITS ORIGINS

So, the Universe needed information generators, certainly at its inception but also at strategic points throughout the history of life on Earth.[27] Despite his commitment to self-organisation, Bernd-Olaf Küppers states: "*There is no complexity generating machine ... that can generate more complexity than is contained in its input.*"[28] Further, DNA information is incompressible, so meaningful information is more like literature than mathematics.[29] This once again implies that it was written by a mind not by an algorithm. As time goes on, fewer and fewer scientists believe that the Universe miraculously created itself out of nothing for no reason; that is anti-scientific. The problem is that it requires an alternative explanation that is believable! John Lennox writes: "*The only source we know of the kind of language type information that we have found in the living cell is mind.*"[30] "*...I believe that injection of information from outside the system is a serious candidate as an inference to the best explanation...*"[31] This principle applies not only to life but to the entire universe.

After decades of study on the origin of life, Professor Emeritus of Biology, Dean Kenyon concluded: "*If science is based on experience, then science tells us that the message encoded in DNA must have originated from an intelligent cause.*"[32] So if God has authored the code for life, then he is not in the gaps in our knowledge but is the origin of the information source code that builds the substance of everything living, as well as the very fabric of the Universe itself. The more we fill in the gaps, the more clearly we see the fingerprints of the Creator. So, science operates in the gaps and God is found in the fabric of the finished product. Lennox writes: "*... the biosphere looks designed because it really was designed;*"[33,34]

9.8 – REVIEWING THE EVIDENCE

The Universe is an extraordinary entity which has been studied by scientists for centuries but our most sophisticated theories, such as the Standard Model of Cosmology and the modern Darwinian synthesis, still struggle to explain even some of the most rudimentary features. These include:

- The relatively sudden appearance of complex entities without explanation (such as the Universe itself, galaxies in the early Cosmos, DNA, life and the Cambrian Explosion.)
- The incompatibility of materialism/determinism with quantum mechanics.[35]
- The 'unreasonable' effectiveness of mathematics.[36]
- The extravagant complexity of life (eg nano-machines, the epigenome – meta-information, redundancy, cellular and organismal information management systems – mRNA and multilayered hormone relays such as the hypothalamic-pituitary-gonadal axis.)[37]
- The vast array of fine-tuning that permits life (such as the charge on an electron, carbon, water and the Goldilocks Zone - amongst one hundred others.)[38]

- The exquisite precision and efficiency of many biological processes (such as the ATP-synthase generator, the citric acid cycle, histone spooling, sexual reproduction and embryogenesis.[39]
- The economy of design evident within the Universe (such as: $E=mc^2$, 25,000 human genes make 100,000 proteins and the processing power of the human mind.)
- The conscious human mind that can exercise reason and is self-aware.
- The origin of meaningful information.

The problem is, as Paul Davies writes: "*Information is not something that is supposed to come for free... Information is something you have to work for. This is really just the second law of thermodynamics revisited, because the spontaneous appearance of information in the universe would be equivalent to a reduction in the entropy of the universe - a violation of the second law, a miracle. Now the fact that the universe contains information is undeniable (because it is not in thermodynamic equilibrium). If information can't get made it must have been there at the beginning, i.e. as part of the initial input. The conclusion we are led to is that the universe came stocked with information or negative entropy, from the word go.*"[40]

Davies points out that as the Universe has progressed, the development of stars and galaxies appear to have contradicted the second law of thermodynamics but concludes that it is the result of gravity: "*Thus gravitationally induced instability is a source of information. ... In some as yet ill-understood way, a huge amount of information evidently lies secreted in the smooth gravitational field of a featureless, uniform gas. As the system evolves, the gas comes out of equilibrium, and information flows from the gravitational field to the matter. Part of this information ends up in the genomes of organisms, as biological information.*"[41] However, an alternative interpretation might be that the change in entropy noted in space-time *is* a measure of the degree of organisation within the Universe but *is not* the biological information that he is seeking to explain; rather, that gravity was merely the *vehicle* that effected the observed galactic formation. I would suggest that the information that effected the change in the cosmic equilibrium associated with the formation of the galaxies lies within the laws of the Universe which, from Figure 7.4 and Table 7.1, lie within Cyberspace which itself is allied with space-time and the space-time fabric and shares similar properties. The laws of the Universe, via the space-time fabric, cause a limited amount of organisation leading to a localised reduction in entropy but they cannot provide the chemical information for the structures to form a DNA molecule, nor the complex information code stored within it. The information within living organisms is semantic or meaningful information which derives from within the quantum domain (rather than from information vehicles such as gravity), so it is *inconceivable* that the information within life derives from the reduction in entropy from gravity in space-time. In light of this new understanding of the relationship between space-time and Cyberspace, it would seem that rather than Cyberspace being an appendage of space-time, as it appeared in Chapter 7, the reverse is true; space-time seems to emerge from Cyberspace (specifically, the space-time fabric) which is in turn empowered by the information fabric within the quantum domain. In a similar way, the information fabric undergirds Mindspace.

Further, as the quantum domain is the origin of the other three domains, it would appear that the three secondary domains originated from within the quantum fabric and were specified by quantum information. Brian Whitworth suggests that as space-time is thought to be a concave surface (due to the accelerating expansion of the Universe), the 'Big Bang' would make more sense if it was a 'Big Bubble' that has progressively expanded from miniscule beginnings at the creation of space-time, from within the quantum fabric. Whitworth writes: "*As the bubble expands, new space comes from the bulk around it.*"[42] Figure 7.4 therefore needs to be adjusted to complete our information map of the Universe, the Infosphere, as depicted in Figure 9.1.

FIGURE 9.1 – INFORMATION MAP OF THE UNIVERSE (THE INFOSPHERE)

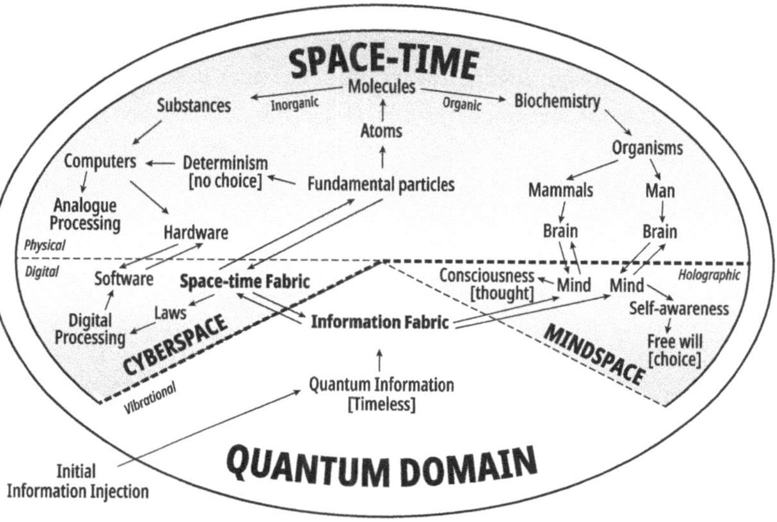

9.9 – CONSTRUCTING THE COSMOS FROM ALGEBRA

So, the Information Model proposes that our material universe is constructed from invisible digital information which at first sight appears unbelievable, so how might this occur? New Scientist investigated this concept in 2025 and details the work of a number of research teams who all seem to be working from mathematics and geometry. An example is the work of mathematicians Jesper Moller Grimstrup and Johannes Aastrup who begin with *"pure algebra"* that builds a *"configuration space"* (as already used in quantum field theory) which they use as their model for space-time, in which they analyse the geometry of the internal space. To their surprize, *"…as they analysed the mathematics, they found the mathematical signatures of fermions encoded within the geometry."* Grimstrup relates: *"By doing this, we relatively easily obtain the building blocks of both general relativity and quantum field theory. So the 'stuff' is there, but not as a part of the foundation…* [further] *Gravity plays no role in our construction in its initial form: gravity in this construction is emergent."*[43]

Then there is the work of mathematician Peter Woit of Columbia University in New York who is using geometrical objects called spinors.

He explains: "*In quantum field theory, spinors are used to describe fermions, the particles of ordinary matter.*" However, spinors derive from geometries in Twistor theory that was first described by Roger Penrose of the University of Oxford in the late 1960s. Twistors, in turn are found in an even more complex geometric entity called an amplituhedron that "*…appears to describe a sizeable chunk of fundamental particle physics.*" Jaroslav Trnka of the University of California Davis elaborates: "'*The amplituhedron approach reformulates the fundamental laws of nature in a geometric language'… Put simply, the amplituhedron's geometry means that what goes in is processed like information in a computer. 'Here the mathematics is fundamental, and physical properties of particle interactions are a consequence of the intricate geometry,' says Trnka.*"[44] It is now over fifty years since Penrose began working on twistors and yet these projects described above are still only works in progress so the fruit is taking time to ripen; even so, with the progress made to date, it does not seems implausible that a mathematical foundation built through increasingly complex geometries already described will be able to model the construction of space-time and the physical entities that we find within it.

CHAPTER TEN

INFORMATION MECHANICS

A NEW UNDERSTANDING

"If this were a purely scientific question and not one that touched on the religious problem, I do not believe that any scientist who examined the evidence would fail to draw the inference that the laws of nuclear physics have been deliberately designed..."

– Fred Hoyle (1957)[1]

Over the past three chapters we have moved from a new classification of information to an information map of the Infosphere to aid our understanding of Information Mechanics that now presents us with an Information Model of the Universe. When we looked at models in Chapter 2, we learnt that they are approximations to enhance our appreciation of the subject being studied, so what can we learn from this Information Model?

10.1 – COUNTERFLOW OF COMPLEXITY

The material universe naturally trends from order to disorder (both physically and informationally) as described by the Second Law of Thermodynamics, also known as the Law of Entropy. We have previously discussed Shannon information which is a measure of the volume of information being sent. Paul Davies relates: *"Shannon's theory hinges on a direct link between information and entropy. ... Shannon's great insight was to spot that noise is a form of disorder, or entropy. By contrast, a signal represents order: compare the carefully arranged dots and dashes of Morse code with the crackle of radio static. In Shannon's theory, information is treated as the opposite of entropy; for that reason, information is sometimes referred to as negative entropy. When information is lost in a noisy communication channel, the entropy rises. ... Communication theory... says that noise destroys information, and that the reverse process, the creation of information by noise, would seem to us a miracle. A message emerging on its own from radio static would be as surprising as the tide making footprints on the beach. We are back with the same old problem: the second law of thermodynamics insists that information can no more spring into being spontaneously than heat can flow from cold to hot."*[2] So, when there is a sustained trend, where the Universe is moving from disorder to greater order or complexity, there must be an input of information and/or energy. This is called the counterflow of complexity and is information dependant.[3] Several examples of information counterflow have already been cited such as the formation of the Universe, the beginning of life and the Cambrian Explosion. Intelligently engineered entities, such as appropriately programmed computers, and life are the only known causes of complexity counterflow.

10.2 – SCIENTIFIC REDUCTIONISM VERSUS THE INTEGRATION OF INFORMATION

Reductionism has been a very powerful tool on our scientific journey as it has progressively allowed scientists to separate our universe into smaller and smaller bits to see how it is made but this process cannot explain all that there is to see, so we need other tools to understand how the pieces mesh together in our immensely complex cosmos. James Shapiro eruditely expressed the situation as: "*We have this terrible dilemma in science. We need to be reductionists to get meaningful results and make observations. But when we take the observations and try to understand what they mean, then we have to stop being reductionists and become integrationists to understand how the things we've identified and singled out fit into the whole picture.*"[4] Reductionism has shown us a myriad of information-driven mechanisms but has not provided a universal understanding of the origin of the information nor of life, not to mention the functioning of mind or how it all fits together. This process of finding the quarks and piece by piece, building a universe is dubbed the 'bottom-up' approach and has worked well for the past four centuries but quantum mechanics has proved significantly more difficult. The other more recent realisation is that living things are made up from a number of interactive parts that, when assembled, provide significantly more benefits than the sum of the individual components alone. A notable example is vision, which emerges when an eye is connected to a functioning brain with the appropriate learning to interpret what is 'seen'. Dissecting out an eye, nerves and a brain won't tell you what sight is. The *intact functioning system* must be studied to understand sight. The living cell is an even more complex example. Michael Behe dubbed this: *irreducible complexity*,[5] where multiple complex elements must be combined for the meta-entity to function at all. The reason for this inability of bottom-up analysis to fully comprehend complex systems is that in the dismantling process, meta-information that produces emergent properties is lost. Lord Jonathan Sacks (1948-2020) commented: "*Science takes things apart to see how they work. Religion puts things together to see what they mean.*"[6] So, a reductionist approach has failed us in these instances and instead, we need to adopt a top-down *systems* approach. However, these are not the only problems. In addition, John Lennox adds: "*...physical reductionism cannot deal with a semantic dimension: words, language, and encoded information.*"[7] As we saw previously, this is vitally important when we consider mind, thought and meaning. However, before we investigate the top-down alternative, it is important to appreciate the difference between mechanism and agency.

10.3 – MECHANISM VERSUS AGENCY

There are different kinds of explanation. We must be careful not to confuse mechanism with agency. *Mechanism* means an information defined process where there is no choice. *Agency* means the ability to choose which information alternative to apply. Choice permits agency that grants the power to effect change. Table 10.1 illustrates various examples of mechanisms with their agencies. A mechanism is a string of instructions and is therefore an information product. These include the four natural laws of the Universe which Paul Davies describes as "*information statements*".[8] Further examples of mechanisms are: computer hardware, biochemistry and the genome in a living organism. Each of these are vehicles that perform a function that controls or permits further events to occur within deterministic space-time. Mechanisms correlate with David Chalmers' category of the *easy problem of consciousness whereas agency is analogous to the hard problem of consciousness.*[9]

When assessing the agency involved with any of these examples, there is implied controlling or meta-information. Computer hardware is driven by software and we now know that the genome is under the direction of the epigenome, while consciousness is associated with mind. However, we do not usually attribute controlling meta-information to the natural laws of the Universe. Edgar Andrews comments that most scientists "*...accept [them] as 'given'...*"[10] We do not know how they work informationally but it seems likely that they are a product of the space-time fabric which is the origin of their determinism.[11] Computer software programmes are known to be the product of mind and hence a derivative of consciousness and thinking. The epigenome is still debated but has more in common with a sophisticated AI programme than an independent mind and I would suggest akin to computer software programming which possesses derived or secondary agency from the mind that created it. The same would also seem to apply to the space-time fabric. Thus, the only true or primary agency that we have discovered in our universe is mind.

TABLE 10.1: EXAMPLES OF MECHANISM AND AGENCY WITHIN THE UNIVERSE

AGENCY [Meta-information]	Space-time Fabric				Software programme	Epigenome	Mind
MECHANISM [Information vehicle]	Gravity	Weak force	Strong force	Electro-magnetism	Computer hardware	Genome	Consciousness
FUNCTIONAL OBJECTIVE [Information outcome]	Macro matter (e.g. planets)	Micro matter (e.g. molecules)	Nano matter (e.g.neutrons)	Charged particles (e.g. electrons)	Information management	Life	Thought

This leads to the obvious question: *Why is mind different?* As mentioned previously, although mind operates within a space-time vehicle, its origin seems to lie in the quantum domain where there appears to be choice due to quantum mechanics. Lennox explores the question of whether living cells have agency, quoting Dennis Noble[12] who attributes the directionality of evolution to the agency of living organisms, while others suggest that the ability of cancers to mutate and avoid our best treatments, as examples of agency. In the case of the former, I would suggest that sentient animals such as birds and mammals do exhibit agency but this is because they have mind. According to Hameroff, it seems likely that a tree or an amoeba would be able to process some limited choices through the pre-programming in their genomes that directs their microtubular networks, but what limited choices they do have certainly does not include the ability to intentionally rewrite their own software to direct their own evolution. Resistant cancers result from mutations in control systems within cells that allow them to career out of control. This is an example of corruption of pre-existing information systems, not agency to choose. Programmed processing is not the same as agency. Only mind has true agency to choose. In the case of plants and microbes, I would argue that the agency lay in the mind of the author of their genetic code.

So, what does this mean for the origins of the natural laws of the Universe and the space-time domain in which they operate? Jonathan Sacks helpfully pointed out that: "*The meaning of a system lies outside the system.*"[13] So when looking for the reason for the existence or origin of a system, looking inside it will not work. This has huge implications for the formation of the Universe. Looking within our universe for its cause will be fruitless but we cannot see outside our universe either to find its cause. Neither can we run experiments to test any creation theories as we are confined to our universe, so science cannot help us here. This means that we must depend on philosophy for plausible explanations but unless there is information from an outside source, we cannot know definitively. What we can deduce is that information is the mechanism and an intelligent mind was the agency. John Lennox suggests that, just as the ancient scriptures have documented, *that mind* is the mind of God.[14,15]

10.4 – TOP-DOWN THINKING

In Figure 4.2, the Universe was described from a reductionist perspective, beginning with space-time and deducing the necessity of an information realm that underpins the quantum domain. However, we have seen that this 'bottom-up' approach[16] leaves us without an explanation for the origin of the vast amount of finely tuned information. So, John Lennox proposes a 'top-down' approach as illustrated in Figure 10.1. It can be seen that this requires an injection or download of information to launch or boot up the system at its formation, which physicists refer to as *initial conditions*.

**FIGURE 10.1:
RELATIONSHIPS WITHIN THE INFOSPHERE**

Systems/Platforms		Mechanisms		Outcomes
Information download [to boot up the system]	→	External source [e.g. the hand (word) of God]	→	Formation/Creation [existence]
↓				
Information Realm [digital]	→	Information Mechanics [platforms]	→	Information Fabric [structure]
↓				
Quantum Domain [probabilistic]	→	Quantum Mechanics [applications]	→	In-determinism [choice/mind]
↓				
Space-time [analogue]	→	Classical Mechanics/ Relativity Theory [algorithms]	→	Determinism [logic/certainty]

Following this line of thinking, once the Universe has its top-down systems established, then bottom-up activity can run the day-to-day events, using the stored information that has been incorporated into the systems. However, John Lennox goes further and suggests that the top-down model was necessary not only for the formation of the Universe at the 'Big Bang' but also that new information must have been infused repeatedly at intervals to account for the appearance of the differing parts of our universe, of which, he suggests a *minimum* of at least two more.[17] This means that the 'Big Bang', which was originally a derisory term coined by Fred Hoyle in 1948, would be more accurately denoted the 'Initial Information Influx' or even the 'Great Information Influx', as at this time the laws of the Universe, the space-time fabric and the appearance of light/energy and hence its derivative alter ego, mass, appear to have arrived. This extraordinary event is referred to by cosmologists as a *singularity*, where there is a discontinuity in the history of the Universe. John Lennox goes on to say: "*...there may be a few singularities additional to the Big Bang that scientists could look for, were they so inclined. The main ones could be the origins of the various types of life including that of plants, fish, birds, animals and humans...* [of particular note are]... *the transition from non-life to life, that is, from the inorganic to the organic. ...* [and] *the transition from animals to humans.... Thus, we have at least three major singularities: at the origin of the universe, at the origin of life, and at the origin of human life. The inputs we are speaking of* *are informational, and, most likely, also energetic, since information does not create on its own, as we have seen with DNA."*[18]

Comparing the top-down thinking of the Information paradigm to the bottom-up thinking of the Material paradigm provides completely different scenarios for the formation of the Universe which are summarised in Table 10.2.

TABLE 10.2: PARADIGMS: MATERIALISM VERSUS INFORMATION

	The Material Paradigm	The Information Paradigm
1	In the beginning there was nothing, then it exploded.	In the beginning there was nothing, until information was introduced to create the information (quantum) realm and fabric.
2	The great explosion somehow created the space-time fabric and the fine-tuning of the laws of the Universe for no reason.	The information injection (through the information fabric) generated the space-time fabric and the laws of the Universe.
3	The random operation of the Universe formed our solar system and the Earth by a statistically impossible accident.	A second injection of information specified the formation of our Earth which is just right for life.
4	Life appeared on Earth, despite the lack of a credible mechanism. [Even Richard Dawkins admits that he cannot explain how the first living cell came into existence.]	A third injection of information provided the genetic coding and specifications for the first cell and life.
5	Life repeatedly became more and more complex with large new chunks of DNA code, despite the Law of Entropy specifying the opposite should occur and Darwinian macro-evolution being incapable of inducting new meaningful information.	A further series of information injections created the Cambrian Explosion, then the sequential appearance of reptiles, birds, mammals and human beings.
Conclusion	This sequence needs unbelievable faith in Materialism-of-the-gaps, as every step has a deficit in at least one of its physics, chemistry, biology and/or logic steps. Yet we are told that it all happened by convenient random happenstance!	This sequence requires an external intelligent mind to write the information code required to specify each successive step. It goes without saying that the code must be phenomenally complex and the mind that devised it, 'all knowing'.

If all this evidence is insufficient to clearly demonstrate the need for a new way to understand our universe, Lennox further points out that our universe has a multi-layered construction, irreducible complexity and is exquisitely finely tuned as well.[19]

10.5 – TESTING THE INFORMATION MODEL OF THE UNIVERSE

So, as we now understand that the Universe began with information as well as energy-matter, we have an alternative way to view the construction and functioning of the Cosmos. An information perspective begins with the origin of information, from which the quantum domain and space-time emerge. This Information Model of the Universe[20] provides greater explanatory power than the Standard Model as it integrates space-time with the quantum domain in a rational way and offers an understanding of consciousness, mind, thought, emotions, sleep and anaesthesia.

Information Mechanics makes the invisible 'visible'. Science is a tool to learn about the Universe and Materialism is one way to use that tool, but we now have others, such as the non-material faculties of mind, which become apparent through an exploration of *impalpable or intangible* things; so why would we not use every means at our disposal to gain understanding?

However, there is a caveat. Karl Popper famously pointed out that a scientific theory can never be ultimately proven but if it is *incorrect*, it should be able to be *falsified*. So, is Information Mechanics falsifiable? The central premise of Information Mechanics is that the entire Universe is underpinned by information, so if it could be shown that information does not underlie the DNA code, the laws of the Universe or quantum mechanics then the theory would either need to be modified or scrapped. At this stage, the evidence does not suggest this is likely but there is much about the laws of the Universe and quantum mechanics that is yet to be understood.

10.6 – SHIFTING TO AN INFORMATION PARADIGM

Materialism and the Standard Model of Physics imply that the Universe was created from nothing, by nothing for no reason, which is why the 'Big Bang' burst brilliantly into being! This sounds absurd and unbelievable because it is! More realistically, the Standard Model is imploding under the weight of evidence from all its inconsistencies and needs to be replaced. As an example of the need to acknowledge the pivotal place of information, Walker, Davies & Ellis ask: *"If information makes a difference in the physical world, which it surely does, then should we not attribute to it causal powers?"*[21]

In 1999, mathematician and co-originator of the Algorithmic Information Theory, Gregory Chaitin asked: *"What if information is primary and matter/energy is a secondary phenomenon?"* to which John Lennox affirms that matter/energy is indeed derivative and therefore secondary.[22] Fact, reason and logic indicate that Materialism, as recognised by the Standard Model of the Universe, cannot be the *whole* story and that something is missing. Space-time provides the *How* that explains the mechanism that leads to knowledge. However, a mechanism is a series of instructions and is bereft of meaning. In contrast, quantum information contributes the *Why* that reveals causation that leads to understanding that furnishes meaning. Further, the *Why* empowers the *How*, but you won't have the full picture until you harmonise the two. This does not mean that space-time is not significant but that it is secondary and hence sub-ordinate in the construction and functioning of the Universe. Information is the elephant in the room. When the Voyager spacecraft beamed back a picture of the Earth from far out in space in 1990, Isaac Asimov (1920-1992) described it as a "pale blue dot" that was used to imply our insignificance in the Universe,[23] however, this presents a purely materialist frame of reference, akin to looking at the Infosphere in Figure 9.1 from above, with space-time in the foreground as the main emphasis, whereas a more realistic approach is to view it from below, placing the quantum domain as the primary focus and space-time as an emergent entity, which is sometimes referred to as *quantum realism.*[24]

The problem is, as Jonathan Sacks previously told us, no system can explain its own origin, as the cause must be outside the system, so every theory of cosmic origins must assume presuppositions. The 'Big Bang' is usually portrayed as not requiring any initial conditions but Brian Whitworth writes: *"The old creation story that God made the universe is now replaced by the story that nothing made everything at a point of infinite density that expanded faster than light by a magical force that then disappeared forever, leaving the universe to sedately evolve into the galaxies, stars and us. Given three miracles, the theory works, but a theory that replaces one miracle by three isn't a very convincing story."*[25] Table 10.3 compares the presuppositions required by our current scientific paradigm versus Information Mechanics. Our answers to these questions determine the way that we interpret the World and informs our understanding of our universe and gives it meaning. Some scientists have argued that the Universe and life have no meaningful purpose but human beings are meaning seeking creatures and Information Mechanics offers us a meaningful understanding.

TABLE 10.3: PRESUPPOSITIONS: CURRENT VERSUS THE INFORMATION PARADIGM

Assumptions that inform the beliefs that we hold.

	Current Scientific Paradigm	Information Mechanics
Origins	In the Standard Model of Cosmology, the initial conditions were self-organising. Assumption: The Cosmos spontaneously generated itself.	As the initial conditions had to be finely tuned for our Cosmos to support life, an originator existed outside it. Assumption: The Cosmos required an architect.
Physics	Materialism explains our entire universe while quantum mechanics is merely theoretical. Assumption: Physical realism explains the Cosmos.	The entire universe is written from information in the language of mathematics. Assumption: there is a non-material foundation that is primary.
Quantum	In the Copenhagen Model, every atom is conscious; in the Many Worlds, every choice makes a new universe. Assumption: realism doesn't matter.	The de Broglie-Bohm Pilot Wave Model provides a more realistic and believable explanation.[26] Assumption: Quantum realism explains the Cosmos.
Life	Life can appear on Earth by a statistically impossible accident. Assumption: The Neo-Darwinian Synthesis explains all life.	Current models cannot explain the origin of the first cell nor the ultra-complexity of the DNA code. Assumption: The code had to be written/designed by an intelligent mind.

It is time to change the paradigm from a purely materialistic way of thinking to an Information Model of the Universe which accounts for both its material *and* non-material characteristics. Switching the paradigm to Information Mechanics is like putting on a new pair of glasses with multifocal lenses. We see the Universe more clearly through an information lens as the Information Model provides significantly greater explanatory power. So, what would a new information paradigm look like? Table 10.4 summarises the changes.

TABLE 10.4:
SUMMARY OF THE INFORMATION PARADIGM

1	Information forms the foundation of the Universe, with the quantum realm as primary and space-time as emergent.
2	Quantum realism points to the rationality of the de Broglie-Bohm model rather than the contrivances required for the Copenhagen & Many Worlds interpretations.
3	The Universe could not have set its own pre-conditions (information parameters), nor written the laws of the Universe – an external mind was required for its precision.
4	Darwinian evolution could not have constructed the first cell, nor written the human DNA code – an intelligent mind was required for the exquisite complexity of life.

The problem is that change can be extremely difficult, and it can be very hard to say goodbye to cherished theories and their comforting familiarity, especially when they have survived for generations and careers have been built upon them. The question is whether the new information requiring an understanding is more valuable than a revered theory that does not provide the explanatory power necessary to answer our queries. The insistence upon only recognising a universe based on materialism alone has led us to stagnate in a 'Shallow Universe Hypothesis', whereas information provides the rich vista of a landscape that connects the quantum and space-time realms that enhances understanding. The next stage in the evolution of science has arrived. Let's embrace it!

10.7 – CONCLUSION

Our frame of reference/level of abstraction is important when we study the Universe; bear in mind that something that is invisible to our human senses does not necessarily mean it is invisible to everything. A submarine deep in the ocean is invisible to the human eye looking down from an aeroplane but not to a whale or dolphin that has sonar capabilities. In the same way, a dog whistle is inaudible to human beings but not to dogs. So, there are circumstances where phenomena are beyond the capabilities of our human senses to detect but they are still real phenomena within our universe. Similarly, just because something is beyond our ability to detect it today, does not mean it does not exist and indeed, that one day we will develop a technique to verify it. A recent example is the 'God particle' or Higgs boson, which was postulated in 1964 but required the CERN linear accelerator to be upgraded to finally detect it in 2012, resulting in the awarding of a Nobel Prize. An earlier example is the positron, predicted by Paul Dirac in 1928 and confirmed by Carl Anderson (1905-1991) in 1932. Then there was the neutrino, postulated by Wolfgang Pauli (1900-1958) in 1930 and 'locked-in' mathematically by Enrico Fermi (1901-1954) in 1934 but not detected physically until 1956. Yet another is the gravitation wave, which Einstein predicted in 1916 but not detected until 2015. These are all entities that logic and science predicted but were not yet detectable. So, unseen entities that are beyond our physical senses are no less real for being invisible to our human faculties. M-theory predicts that seven dimensions in our Universe are real but currently unable to be detected. So, we need to keep an open mind and the humility to acknowledge that there may still be much to be discovered in our universe. With this understanding, where mathematics and/or logic implies or predicts that there are hidden parts to our universe, we need to take a rigorous approach to investigate any hypothesis that may explain discontinuities or irregularities, as such research may bear much fruit.

John Lennox concludes: *"I think the time has come to accept both the existence of an 'information first' universe and of additional historical informational singularities - at least at the origin of life and the origin of human consciousness. However, information first essentially implies consciousness first - not, of course, human conscious, but the consciousness of the Mind of God. Perhaps we should re-calibrate and say, with Nobel laureate for physics, Max Planck (1858- 1947): I regard consciousness as fundamental. I regard matter as derivative from consciousness."*[27] If we accept Professor Lennox's words, then the Universe has a top-down codification. This top-down causation means that the Universe must have been front-loaded with information as discussed previously and therefore that information is the fundamental and primary entity in our universe. From this information the quantum world is derived, while the materialistic domain emerges from the quantum (as Stapp described in Chapter 6); so we need to flip what we have traditionally regarded as primary and secondary. Quantum is formative while space-time is derivative.[28] Information in the quantum domain is in a state of superposition and remains multi-potent or versatile but when it encounters space-time, its waveform collapses and it becomes committed or fixed information (materialistic and hence deterministic).

No less significantly, to be effective, information needs to be managed. It doesn't meaningfully manage itself. This requires control software or meta-information, like the epigenome or an AI programme that manages a computer system. Yet this complex control software does not self-assemble; it too needs to be written, which means it requires design to exist. This is because inducting meaningful information into any system is an active process that requires mind and never occurs passively or by chance.[29] For the materialist, the ultimate nature of reality is mass/energy, which produces mechanisms but lacks agency and we previously saw that mechanism and agency are not the same thing. Space-time, the quantum world and information are *all* vitally important for our understanding of the Universe but none of them has been shown individually to create themselves nor derive the rest of the Cosmos. That requires a master programmer, a mind of unimaginable capacity to write the algorithms which paint the landscapes that construct, run and decorate our world.[30] However, we have already seen that this Creator must be *external* to the creation.[31] For Aristotle and Aquinas, this is the 'unmoved mover'.[32,33]

From the foregoing, it is patently clear that there is overwhelming evidence that we live in an exquisitely complex universe that is both specified by, and runs on, defined information. So, going forward, to have a complete understanding of our existence in the Universe, both our space-time existence and our quantum reality, must be considered from an information perspective. For many of us, this will require a radical rethink as space-time and the quantum domain often appear to be in tension, if not completely contradictory, but without it, we cannot contemplate our whole way of life. Even though the mechanisms are intricate, the concept of Information Mechanics is profoundly simple:

a. *information generates/activates thoughts or 'software' via the quantum domain;*
b. *thoughts and/or 'software' generate actions in space-time.*

Further, meaningful information not only has explanatory power but it also has causal power. Therefore, it is more than an entity – it has agency![27] Materialism (in space-time) describes the mechanisms but information is the agent. Another way of saying this is that information in the quantum realm creates the message, while space-time provides its materialistic vehicle that permits it to happen. In 1847, William Whewell, who had coined the term 'scientist' in 1834, wrote: "*The facts are known but they are insulated and unconnected... The pearls are there but they will not hang together until some one provides the string.*"[34] It is now clear that information is that string! It is time to acknowledge that our *governing paradigm* is not merely materialistic but *informational*, that *incorporates* material space-time. Simply acknowledging space-time without recognising the activity of the underlying quantum domain and ignoring the role of information leads to a very shallow understanding of our universe.

So, in the light of this new understanding provided by the Information Model, it is timely to also begin measuring *our* significance as human beings not in terms of space-time and Materialism but in terms of Mindspace and the quantum domain; human beings are firstly 'mind' and secondly 'body'. Let us open our eyes to see the fullness of everything that is there. In acknowledging that, however, the ultimate agent is not the information itself, but that it is a tool in the hand of the author of the code, the mind *behind* the cypher. Quantum physicist and Nobel laureate, Werner Heisenberg (1901-1976) is reported to have stated: "*The first gulp from the glass of natural science will turn you into an atheist but at the bottom of the glass God is waiting for you.*"[35] Materialists will undoubtedly report that in their experience, when they have reached the bottom of their glass, they have never seen God's face peering up at them, however, a closer inspection will reveal the molecular structure of the glass that is built from quantum components, underwritten by an information code that is undoubtedly his digital signature.

ACKNOWLEDGMENTS

I have been actively working on the ideas in this book for twenty years but I realize that the motivation to pursue this topic was seeded much earlier by my parents and my teachers who taught me to think about our universe and I have always been fascinated by life. My father showed me how to take a few drops of creek water and look at them under his microscope to see the life teeming within and to ask how and why? My mother patiently encouraged me to persevere in the face of disappointment and to try again when I failed. My headmaster, Alistair Mackerras encouraged the students to love learning and to aspire to be all-rounders or today we would say, generalists, who have a wide scope of skills, knowledge and understanding.

In particular, fifty years ago in our final year at school our physics teacher, Mr Kerry Crowe-Mai pointed to the disparity between Relativity Theory and Quantum Mechanics and challenged the class to solve the mystery which obviously had a solution because the Universe functions seamlessly and upon which I have pondered over the intervening years.

More recently, Dr Ken Meyer has encouraged me to maintain rigor in my lines of argument while Dr Mark Glasgow has challenged me to read more widely than felt comfortable, to challenge ideas that were becoming too familiar and unhelpful. Meanwhile, Dr David Learoyd listened patiently to my evolving ideas, helping me to refine woolly concepts and giving inciteful feedback that extended the models.

Gary and Anne Marie Ezzo have continued to encourage, assisted with the structure and design of the book, as well as the production, while my children and daughters-in-law have given constructive feedback to develop the cover, content and presentation.

Finally, I wish to thank my wife, Dawn, who has patiently listened to the numerous reiterations of the narratives and steadfastly supported this endeavour through endless encouragement over the past six years.

NOTES

Preface

1. Lowe, I.: Emeritus Professor of Science, Technology and Society at Griffith Univ., Qld.
2. Whewell, W.: *The Philosophy of the Inductive Sciences*, John W. Parker, London (1847) Vol.2, p.36 *in* McGrath, A.E.: *Surprised by Meaning: Science, Faith and How We Make Sense of Things*. Westminster John Knox Press, Kentucky (2011) p.2.
3. Gibson, W.: Burning Chrome. *Omni* Magazine, 72 Jul (1982) p.46.
4. Davies, P. & Gregersen, N.H.: (eds) *Information and the Nature of Reality*. Cambridge Univ. Press (2010) p.3.
5. Floridi, L.: *Information: A Very Short Introduction*. Oxford Univ. Press (2010) pp.110-1.
6. Davies, P.: Universe from bit. *in* Davies, P. & Gregersen, N.H. (eds) *Information and the Nature of Reality*. Cambridge Univ. Press (2010) p.75.
7. For further discussion of paradoxes see: Wiles, R.B.: *The Information Paradox*. Information Press Australasia (2014).
8. Walker, S.I., Davies, P.C.W. & Ellis, G.F.R.: *From Matter to Life: Information and Causality*. Cambridge Univ. Press (2017) p.1.
9. Lennox, J.C.: *Cosmic Chemistry: Do God and Science mix?* Lion Books, UK (2021) p.115.
10. McGrath, A.E.: *Natural Philosophy: On Retrieving a Lost Disciplinary Imaginary*. Oxford Univ. Press (2023) p.94.

Chapter 1 – Information Mechanics: Why We Need It

1. Floridi, L.: *The Philosophy of Information*. Oxford Univ. Press (2011) p.361.
2. Lichtman, J.: Harvard lecture in Schilling, D.R.:'Knowledge Doubling Every 12 Months, Soon to be Every 12 Hours' in *Industry Tap* 19 Apr (2013).
3. Lennox, J.C.: *Cosmic Chemistry: Do God and Science Mix?* Lion Books, UK (2021) pp.31-32.
4. Stapp, H.P.:'Quantum Theory and Free Will.' Nature Cham, Switzerland, Springer (2017) p.9 in Lennox, J.C.: *Cosmic Chemistry*. Lion Books, UK (2021) p.361.
5. Floridi, L.: *The Philosophy of Information*. Oxford Univ. Press (2011) p.4.
6. Kantor, F.W.: *A brief introduction to information mechanics* (PDF) (1994). https://web.archive.org/web/20160615102231/https://www.gravityresearchfoundation.org/pdf/awarded/1974/kantor.pdf [Accessed 19 Sep 2024]
7. Zuse, K.E.O.: *Calculating Space - Translation of: Rechnender Raum*. (PDF) Feb (1970) MIT Technical Translation by Aztec School of Languages, Inc. Cambridge, Ma, USA [Accessed 19 Sep 2024] https://web.archive.org/web/20200325103231/ftp://ftp.idsia.ch/pub/juergen/zuserechnenderraum.pdf
8. Information about Edward Fredkin. https://web.archive.org/web/20131029191129/http://english.turkcebilgi.com/Edward+Fredkin & http://www.digitalphilosophy.org/ [Accessed 19 Sep 2024]
9. Wheeler, J.A.:'Information, Physics, Quantum: The Search for Links' *in Proceedings of the Third International Symposium on the Foundations of Quantum Mechanics held in Tokyo, 1989* (1990).
10. Floridi, L.: *The Philosophy of Information*. Oxford Univ. Press (2011) p.1.

11. Nyquist, H.: *Bell System Technical Journal*, Vol. 3, Issue 2, (1924) pp324-346.

12. Shannon, C.E. & Weaver, W.: *The Mathematical Theory of Communication*. Univ. of Illinois Press (1949).

13. Davies, P. & Gregersen, N.H.: (eds) *Information and the Nature of Reality*. Cambridge Univ. Press (2010) p.4.

14. Floridi, L.: *Information: A Very Short Introduction*. Oxford Univ. Press (2010) p.45.

15. Davies, P.: *The Fifth Miracle: The Search for the Origin of Life*. Penguin Books (1999) p.242.

16. Some computers are being programmed to analyse information in defined situations such as flagging abnormal blood test results for a clinician to review but they cannot write new software routines for themselves as they problem-solve on the run and appreciating meaning is also far beyond their current capability.

17. Bekenstein, J.D.: 'Information in the Holographic Universe', *Scientific American*, 289(2) (2003) p.59.

18. Lloyd, S.: 'Life, the Universe, and Everything.' *Wired Magazine* 1st Mar (2006).

19. Gleick, J.: *The Information: A History, A Theory, A Flood*. Pantheon, NY p.9f (2011) in Glattfelder, J.B.: Information - Consciousness – Reality. Springer (2019) p.479.

20. Wheeler, J.A. 'Information, Physics, Quantum: The Search for Links' (1990) in Zurek, W.H.: (ed) *Complexity, Entropy, and the Physics of Information*, Addison-Wesley, Redwood City, 354-368 https://philpapers.org/archive/WHEIPQ.pdf (Accessed 28/10/2023).

21. Einstein, A.: *Letters to Solovine: 1906-1955*, Philosophical Library, NY (1987) p.131; for further discussion see Wiles, R.B.: The Information Paradox Information Press, Australasia (2014) p.272.

22. Wigner, E.P.: 'The unreasonable effectiveness of mathematics.' *Communications in Pure and Applied Mathematics*, 13 (1960) pp.1-14; see Lennox, J.C.: God's Undertaker: Has Science Buried God? Lion Books, Oxford (2007) p.60.

23. Stachel, J. et al.: (eds) Einstein, A.: 'The 1905 Papers' in *The Collected Papers of Albert Einstein*. Princeton University Press, NJ (1987).

24. de Duve, C.: *Life Evolving*. Oxford Univ. Press, NY (2002) p.199; for further discussion see Wiles, R.B.: The Information Paradox (2014) p.105.

25. Alberts, B.: The Cell as a collection of Protein Machines. *Cell* 92:291 (1988); for further discussion see Wiles, R.B.: *The Information Paradox* (2014) p.19.

26. Chalmers, D.: *The Conscious Mind*. Oxford Univ. Press, Oxford (1996) p.168.

27. Rees, M.: *Just Six Numbers: The Deep Forces that Shape the Universe*. Weidenfield & Nicholson, London (1999); for further discussion see Wiles, RB: The Information Paradox (2014) pp.192-193.

28. Floridi, L.: *The Philosophy of Information*. Oxford Univ. Press (2011).

29. Pearcey, N.: *Total Truth: Liberating Christianity from its Cultural Captivity*. Crossway, Illinois (2005) p.220.

30. This proposed new paradigm of Information Mechanics differs from Frederick Kantor's concepts from 1974. (See Chapter 1, p.19).

31. Kuhn, T.S.: *The Structure of Scientific Revolutions*. Univ. of Chicago Press, Chicago (1962).

32. Hawking, S. & Mlodenow, L.: *The Grand Design*. Bantam Press, UK (2010) p.115.

33. Hawking, S. & Mlodenow, L.: *The Grand Design*. Bantam Press, UK (2010) p.117.

34. Rovelli, C.: *Reality Is Not What It Seems: The Journey to Quantum Gravity*. Allen Lane (2016) p.137.

35. Jelonesi, J.: 'Carlo Rovelli & Tibor Molnar - An answer for Carlo Rovelli and his quantum problem.' *The Philosopher's Zone* – ABC Radio, Aust. 2 Jul (2017). https://www.abc.net.au/listen/programs/philosopherszone/an-answer-for-carlo-rovelli-and-his-quantum-problem/8659546 (Accessed 29/9/2024)
36. Lennox, J.C.: *Cosmic Chemistry*. Lion Books, UK (2021) p.330.
37. Floridi, L.: *The Philosophy of Information*. Oxford Univ. Press (2011) p.39.
38. See Armstrong, D.M.: *A Materialist Theory of the Mind*. Routledge (1968).
39. For further discussion and the effect of brain pathologies, such as strokes, dementia etc, see: Wiles, R.B. *The Information Paradox* Information Press (2014) pp.311-312.

Chapter 2 – Exploring Information

1. Lloyd, S.: *Programming the Universe*. Knopf, NY (2006) p.40 *in* Gregersen, N.H.: 'God, Matter and Information: towards a Stoicizing Logos Christology' *in* Davies, P. & Gregersen, N.H. (eds) *Information and the Nature of Reality*. Cambridge Univ. Press (2010) p.334.
2. Jelonesi, J.: 'Carlo Rovelli & Tibor Molnar - An answer for Carlo Rovelli and his quantum problem.' *The Philosopher's Zone* – ABC Radio, Aust. 2 Jul (2017). https://www.abc.net.au/listen/programs/philosopherszone/an-answer-for-carlo-rovelli-and-his-quantum-problem/8659546 (Accessed 29/9/2024)
3. For further discussion regarding fundamental entities see: Lennox, J.C.: *Cosmic Chemistry*. Lion Books, UK (2021) pp.318-320.
4. Floridi, L.: *The Philosophy of Information*. Oxford Univ. Press (2011) p.41.
5. Hawking, S. & Mlodenow, L.: *The Grand Design*. Bantam Press, UK (2010) pp.42-43.
6. Krakauer, D.: 'Cryptographic Nature' *in* Walker, S.I., Davies, P.C.W. & Ellis, G.F.R.: *From Matter to Life: Information and Causality*. Cambridge Univ. Press (2017) p.167.
7. For a discussion on the materiality/physicality of information see Chapter 4.
8. Dirckx, S.: *Am I just my brain*. The Good Book Co. UK (2019) pp.26-27 & 85.
9. Lennox, J.C.: *God's Undertaker*. Lion Books, Oxford (2007) p.67.
10. Lennox, J.C.: *God's Undertaker*. Lion Books, Oxford (2007) p.183, Ref. 31.
11. Strobel, L.: *The Case for a Creator*. Zondervan (2004) p.101.
12. Lennox, J.C.: *Cosmic Chemistry*. Lion Books, UK (2021) pp.319-320.
13. Floridi, L.: *The Philosophy of Information*. Oxford Univ. Press (2011) p.30.
14. Floridi, L.: *The Philosophy of Information*. Oxford Univ. Press (2011) p.85.
15. Floridi, L.: *The Philosophy of Information*. Oxford Univ. Press (2011) p.82.
16. A datum is a point of difference in the Universe. (See Ref. 14 above).
17. Lennox, J.C.: *Cosmic Chemistry*. Lion Books, London (2021) p.174.
18. Marks, R.J. II, Dembski, W.A. & Ewart, W.: *Introduction to Evolutionary Informatics*. World Scientific Publishing Co. Singapore (2017) p.282.
19. Floridi, L.: *The Philosophy of Information*. Oxford Univ. Press (2011) p.39.
20. Wiles, R.B.: *The Information Paradox* Information Press (2014) pp.309-310.

Chapter 3 – The Characteristics of Information

1. Weiner, N.: *Cybernetics or Control and Communication in the Animal and the Machine.* 2nd Ed. MIT Press, Cambridge, Ma. (1948) p.132.
2. Davies, P.: 'Life Force.' *New Scientist* 163 (2204):27-30 (1999).
3. Bekenstein, J.D.: 'Information in the Holographic Universe.' *Scientific American* 289(2) (2003) pp.59-65.
4. Sheppard, R.Z.: 'Rock Candy.' *Time Magazine* 12 April (1971).
5. Floridi, L. *Information: A Very Short Introduction.* Oxford Univ. Press (2010) p.9.
6. Walker, S.I., Davies, P.C.W. & Ellis, G.F.R.: *From Matter to Life: Information and Causality.* Cambridge Univ. Press (2017) p.1.
7. Pearcey, N.: *Total Truth.* Crossway, Wheaton, Il. (2005) p.197.
8. Floridi, L.: *The Philosophy of Information.* Oxford Univ. Press (2011) p.87.
9. Lennox, J.C.: *Cosmic Chemistry: Do God and Science mix?* Lion Books, London (2021) p.189.
10. Floridi, L.: *The Philosophy of Information.* Oxford Univ. Press (2011) p.34.
11. Lennox, J.C.: *Cosmic Chemistry.* Lion Books, London (2021) p.324.
12. Floridi, L: *The Philosophy of Information.* Oxford Univ. Press (2011) p.202.
13. For further discussion see: Wiles, R.B.: *The Information Paradox* Information Press, Australasia (2014) p.18.
14. Einstein, A.: *Letters to Solovine: 1906-1955.* Philosophical Library, NY (1987) p.131 *in* Lennox, J.C.: *God's Undertaker: Has Science Buried God?* Lion Books, Oxford (2007) p.59.
15. Davies, P. & Gribbin, J.: *The Matter Myth,* Ch. 1 'The Death of Materialism.' Simon & Shuster, NY (1992) in Lennox, J.C.: *Cosmic Chemistry.* Lion Books, London (2021) p.362.

Chapter 4 – Surprizing Outcomes

1. Davies, P.: 'Life Force.' *New Scientist* 163 (2204):27-30 (1999).
2. Rovelli, C.: *Reality Is Not What It Seems: The Journey to Quantum Gravity.* Allen Lane (2016) p.28.
3. Drake, S.: *Discoveries and Opinions of Galileo.* Doubleday-Anchor, NY (1957) in Davies, P. & Gregersen, N.H.: (eds) *Information and the Nature of Reality.* Cambridge Univ. Press (2010) p.67.
4. Lennox, J.C.: *Cosmic Chemistry.* Lion Books, London (2021) p.74.
5. O'Connell, C.: 'The Mystery of the Missing Antimatter.' *Cosmos* 69 (2016) p.36.
6. Gleick, J.: *The Information: A History, A Theory, a Flood.* Pantheon, NY p.357 (2011) in Glattfelder, J.B.: *Information - Consciousness – Reality.* Springer (2019) p.480.
7. Moskowitz, C.: 'Tangled up in Spacetime.' *Scientific American* 26 Oct (2016).
8. Rovelli, C.: *Reality Is Not What It Seems.* Allen Lane (2016) p.102.
9. Rees, M.: *Just Six Numbers: The Deep Forces that Shape the Universe.* Weidenfield & Nicholson, London (1999).
10. Lennox, J.C.: *Cosmic Chemistry.* Lion Books, London (2021) p.149.
11. Clark, S.: 'Unknown Earth: How come Earth got all the good Stuff?' *New Scientist* 199(2675): 29-30 (2008).

12. Davies, P.: 'Life, the universe and everything.' *Cosmos* 14 (2007).

13. Hoyle, F.: *Annual Review of Astronomy and Astrophysics*, 20, 1982 p.16 in Lennox, J.C.: Cosmic Chemistry. Lion Books, London (2021) p148.

14. Bradley, W.L.: 'Nature. Design or Designoid' in Dembski, W.A.: (ed) *Mere Creation. Science, Faith and Intelligent Design.* InterVarsity Press (1998) p.40.

15. Davies, P.: *The Cosmic Blueprint.* Simon and Shuster, NY (1988) p.203.

16. Landauer, R. et al: 'Information is physical.' *Physics Today* 44(5) pp.23-29 (1991); Landauer, R.: 'The Physical nature of information.' *Physics Letters* A, 217(4-5) pp.188-193 (1996).

17. Lennox, J.C.: *Cosmic Chemistry.* Lion Books, London (2021) p.331.

18. Lennox, J.C.: *Cosmic Chemistry.* Lion Books, London (2021) p.330.

19. Timpson, C.G.: *Quantum Information Theory and the Foundations of Quantum Mechanics.* Oxford Univ. Press, Oxford (2010) in Lennox, J.C.: Cosmic Chemistry (2021) p.330.

20. Williams G.C.: 'A Package of Information' in Brockman, J.: *The Third Culture: Beyond the Scientific Revolution.* Simon & Schuster, NY (1995) p.43 in Pearcey, N.: Total Truth. Crossway, Wheaton, Il. (2005) p.197.

21. Armstrong, D.M.: *A Materialist Theory of Mind.* Routledge (1968).

22. Searle, J.R.: 'Biological Naturalism' in *The Oxford Companion to Philosophy* (2nd ed.) Oxford Univ. Press (2005).

23. Lloyd, S.: 'The Computational Universe' *in* Davies, P. & Gregersen, N.H.: *Information and the Nature of Reality.* Cambridge Univ. Press (2014) pp. 97-98.

24. Küppers, B-O: 'Information and communication in living matter' in Davies, P. & Gregersen, N.H.: *Information and the Nature of Reality* (2010) pp.180-1.

25. Lennox, J.C.: *Cosmic Chemistry.* Lion Books, London (2021) p.205.

26. Braunstein, S.L., Pati, A.K.: 'Quantum Information Cannot Be Completely Hidden in Correlations: Implications for the Black-Hole Information Paradox.' *Physical Review Letters.* 98 (8): 080502 (2007).

27. Samal, J.R.; Pati, A.K.; Kumar, A.: 'Experimental Test of the Quantum No-Hiding Theorem.' *Physical Review Letters* 106 (8): 080401 (2011).

28. For further discussion on the conservation of information see: Lennox, J.C.: *Cosmic Chemistry.* Lion Books, London (2021) p.203.

29. Glattfelder, J.B.: *Information - Consciousness – Reality.* Springer (2019) p.498.

30. Floridi, L.: *The Philosophy of Information.* Oxford Univ. Press (2011) p.15.

31. Dawkins, R.: Public debate in Oxford on 20th Jan. 1999 in Lennox, J.C.: *Cosmic Chemistry.* Lion Books, London (2021) p.195.

32. Gijbers, A.: The Dialogue between Neuroscience and Theology. *Luke's Journal* 11(2): 8-9 (2006).

33. Lennox, J.C.: *Cosmic Chemistry.* Lion Books, London (2021) p.195.

34. Andrews, E.H.: *Who Made God? Searching for a Theory of Everything.* EP Books, Darlington, UK (2009) pp.150-152.

35. For further discussion on the formation of the natural laws of the universe and the problems with a self-generating universe see: Wiles, R.B.: *The Mind in the Matrix – What the Complexity of the Universe tells us about Meaning.* Information Press Australasia pp.34-38.

36. Lennox, J.C.: *Cosmic Chemistry.* Lion Books, London (2021) p.333.

Chapter 5 – Building Complexity

1. Davies, P.: 'Life Force.' *New Scientist* 163 (2204):27-30 (1999).
2. Walker, S.I., Davies, P.C.W. & Ellis, G.F.R.: *From Matter to Life: Information and Causality.* Cambridge Univ. Press (2017) p.5.
3. Davies, P.: *The Fifth Miracle: The Search for the origin of Life.* Penguin Books (1999) p.93.
4. Walker, S.I., Davies, P.C.W.: 'The algorithmic origins of life.' *J R Soc* Interface 10: 20120869 (2013).
5. Note that the DNA code within the genome is a digital store while the metabolic activity in the cell is generally processing analogue information.
6. Walker, S.I. & Davies, P.C.W.: 'The Hard Problem of Life.' in Walker, S.I., Davies, P.C.W. & Ellis, G.F.R.: *From Matter to Life: Information and Causality.* Cambridge Univ. Press (2017) p.23.
7. Schwartz, J., Stapp, H. and Beauregard, M.: 'Quantum physics in neuroscience and psychology: a neurophysical model of mind/brain interaction.' *Philosophical Transactions of the Royal Society,* B: 360:1458 (2005) pp.1309-27 in Lennox, J.C.: *Cosmic Chemistry: Do God and Science mix?* Lion Books, UK (2021) p.363.
8. Küppers, B-O.: 'Information and the Origin of Life.' Cambridge, MA: the MIT Press (1990) *in* 'Information and Communication in Living Matter.' *in* Davies, P. & Gregersen, N.: (eds) *Information and the Nature of Reality: from Physics to Metaphysics.* Cambridge Univ. Press (2010) p.170.
9. Lennox, J.C.: *Cosmic Chemistry.* Lion Books, UK (2021) p.203.
10. Lennox, J.C.: *Cosmic Chemistry.* (2021) p.175.
11. Behe, M.J.: *Darwin's Black Box: The Biochemical Challenge to Evolution.* Free Press, NY (1996).
12. See: Wiles, R.B.: *The Information Paradox.* Information Press (2014) p.228.
13. Gould, S.J.: – review of Mark C. Taylor's *The Moment of Complexity: Emerging Network Culture* in *The London Review of Books* 357(1), 2011, 22 Feb (2002) p.5 *in* Lennox, J.C.: *Cosmic Chemistry* (2021) p.276.
14. Davies, P.: *The Fifth Miracle* (1999) p.104.
15. Noble, D.: *Replace the Modern Synthesis (Neo-Darwinism): An Interview with Dennis Noble* – 5 Apr (2014) in Lennox, J.C.: Cosmic Chemistry (2021) p.305.
16. See: Walker, S.I., Davies, P.C.W.: 'The algorithmic origins of life.' *J R Soc* Interface 10: 20120869 (2013) p.5.
17. Noble, D.: *The Music of Life: Biology Beyond the Genome.* Oxford Univ. Press, Oxford (2006) p.41.
18. Lennox, J.C.: *Cosmic Chemistry* Lion Books, UK (2021) p.329.
19. See Hofstadter, D.: *Gödel, Escher, Bach: An Eternal Golden Braid.* Penguin, London (1979) p.548 in Lennox, J.C.: Cosmic Chemistry (2021) p.185.
20. Adami, C. & LaBar, T.: 'From Entropy to Information: Biased Typewriters and the Origin of Life.' in Walker, S.I., Davies, P.C.W. & Ellis, G.F.R.: *From Matter to Life: Information and Causality.* Cambridge Univ. Press (2017) pp.147-148.
21. Hamlet has 29551 words x ~6 characters/word + 1 space = 206,857.
22. Dawkins, R.: *Climbing Mount Improbable.* Penguin. London (2006) p.68.
23. For further discussion see: Wiles, R.B.: *The Information Paradox* (2014) pp.147-152.
24. For further information on Complexity Theory and the origin of life see: Wiles, R.B.: *The Information Paradox* (2014) p.148.

25. See Fuller, S.: *Science Vs. Religion.* Polity, Cambridge (2007) p.89 in Lennox, J.C.: *Cosmic Chemistry* (2021) p.287.
26. Davies, P.: *The Fifth Miracle* (1999) p.122.
27. Müller, G.B.:'Homology: The Evolution of Morphological Organisation' *in* Müller, G.B. & Newman, S.A.: (eds) *Origination of Organismal form. Beyond the Gene in Developmental and Evolutionary Biology.* MIT Press, Harvard, MA. Vienna Series in Theoretical Biology, (2003) p.51 in Lennox, J.C.: *Cosmic Chemistry* (2021) p.301.
28. Behe, M.J.: *The Edge of Evolution.* Simon and Schuster, NY (2007) pp.55-63.
29. Lennox, J.C.: *Cosmic Chemistry* (2021) p.198.
30. Lennox, J.C.: *Cosmic Chemistry* (2021) p.185.
31. For further discussion see: Lennox, J.C.: *Cosmic Chemistry* (2021) p.351.
32. Lennox, J.C.: *Cosmic Chemistry* (2021) p.300.
33. Lennox, J.C.: *Cosmic Chemistry* (2021) p.206.

Chapter 6 – The Mind Conundrum

1. *What is Life?* Daniel Sander Hoffmann im Gespräch mit Bernd-Olaf Küppers. Episteme 124 (2002).
2. Lennox, J.C.: *Cosmic Chemistry* Lion Books, UK (2021) p.351.
3. Wertheim, M.: Alas Poor Zombie. *Cosmos* 11:86-91 (2006).
4. Dunbar, R.: Why are Humans not Just great Apes? *in* Pasternak, C.: (ed) *What makes us human?* One World Publications, Oxford (2007) pp.37-48 *in* Cosmos 20:47-49 (2008).
5. For further discussion see: Wiles, R.B.: *The Information Paradox.* Information Press (2014) pp.301-305.
6. Levine, J.: Materialism and Qualia: The Explanatory Gap. *Pacific Philosophical Quarterly* 64 (1983) 354-361.
7. Chalmers, D.: *The Conscious Mind.* Oxford Univ. Press, Oxford (1996) *in* Lennox, J.C.: *Cosmic Chemistry* (2021) p.355.
8. Kerskens, C.M. & Pérez, D.L.: Experimental indications of non-classical brain functions. *J. of Physics Communications* (2022) 6(10):105001.
9. Hameroff, S., Penrose, R.: Consciousness in the universe: A review of the 'Orch OR' theory. *Phys. Life Rev.* 11:39–78 (2014).
10. Musser, G.: Can quantum hints in the brain revive a radical consciousness theory? *New Scientist* https://www.newscientist.com › article › mg26134740-8. 17 Jan (2024).
11. Hameroff, S.: Microtubules: The Gateway to Consciousness. Mindfest Conference, Florida 27 Mar 2024 https://www.youtube.com/watch?v=0_bQwdJir1o Accessed: 24 Apr 2025.
12. Dennett, D.C.: *Consciousness Explained.* Little, Brown & Company. (1991).
13. Stenger, V.: The Myth of Quantum Consciousness. (PDF). *The Humanist.* Vol. 53, no. 3 (1992) pp.13–15.
14. Wigner, E.: Remarks on the Mind-Body Question. in Good, I. J.: (ed) *Philosophical Reflections and Syntheses.* The Collected Works of Eugene Paul Wigner. London: Heinemann (1995) pp.284–302.
15. Discovery of Quantum Vibrations in 'Microtubules' inside Brain Neurons Corroborates Controversial 20-Year-Old Theory of Consciousness. Elsevier. 16 January 2014.

16. Jibu, M., Yasue, K.: What is mind? Quantum field theory of evanescent photons in brain as quantum theory of consciousness. *Informatica.* 21:471–490 (1997).

17. Rourk, C.J.: Ferritin and neuromelanin 'quantum dot' array structures in dopamine neurons of the substantia nigra pars compacta and norepinephrine neurons of the locus coeruleus. *Biosystems* 171:48 (2018).

18. Such as the sinoatrial and atrioventricular nodes in the heart.

19. Nishiyama, A. et al.: Holographic Brain Theory: Super-Radiance, Memory Capacity and Control Theory. *Int J Mol Sci.* 25(4):2399 (2024).

20. Babcock, N.S. et al.: Ultraviolet Superradiance from Mega-Networks of Tryptophan in Biological Architectures. *J. of Phys.* Chem. B. Vol 128/Issue 17. April 19, 2024.

21. Khan, S. et al.: Microtubule-Stabilizer Epothilone B Delays Anesthetic-Induced Unconsciousness in Rats eNeuro 15 August 2024, 11(8).

22. For a YouTube presentation of this see: *Was Penrose Right? NEW EVIDENCE for Quantum Effects in the Brain.* PBS Spacetime (26 Jul 2024) https://www.youtube.com/watch?v=xa2Kpkksf3k

23. Hoffman, D.: We assume our senses see reality as it is, but that could be just an illusion of evolution. in Cossins, D. (ed): How to Think about Reality. *New Scientist* England (2025) p.81.

24. Tegmark, M.: Importance of quantum decoherence in brain processes. *Physical Review E.* 61(4). (2000) pp.4194–4206. However, the relatively slow response times of the macroscopic neurons is the precise reason that a quantum explanation is required. Penrose points out that ping pong would be a very slow affair if the players' response times were pegged to the macroscopic neuronal firing rate. See: Eagleman, D.: Inner Cosmos: Ep110: *Is Consciousness related to quantum physics?* (2nd July 2025) https://www.youtube.com/watch?v=nok4GhijvAA

25. Hameroff, S.: Is Consciousness a Quantum Orchestra? Microtubules, Objective Reduction & ORCH-OR. *Mind-body Solution podcast.* 24 Jan 2025. www.youtube.com/watch?v=6iCzHwL2EY0

26. Whitworth, B.: *Quantum Realism.* Chapter 6, pp.28-32 (2025). https://brianwhitworth.com/quantum-realism/

27. Craddock, T.J.A., Tuszynski, J.A. & Hameroff, S.: Cytoskeletal Signaling: Is Memory Encoded in Microtubule Lattices by CaMKII Phosphorylation? *PLoS Computational Biology,* Vol.8:3 (2012).

28. Nagle, T.: *Mind and Cosmos.* Oxford Univ. Press, Oxford (2012) pp.14-15.

29. Schwartz, J., Stapp, H. & Beauregard, M.: Quantum physics in neuroscience and psychology: a neurophysical model of mind/brain interaction. *Philosophical Transactions of the Royal Society,* B: 360:1458 (2005) pp.1309-27 in Lennox, J.C.: Cosmic Chemistry (2021) p.363.

30. Lennox, J.C.: *Cosmic Chemistry* (2021) p.364.

31. Lennox, J.C.: *Cosmic Chemistry* (2021) p.361.

32. Brooks, R.: Professor of Evolutionary Biology, Univ. of NSW, Sydney. *Sydney Morning Herald:* News Review (Sep 24-25, 2011) p.18.

33. Stapp, H.P.: *Quantum Theory and Free Will.* Springer Nature: Cham, Switzerland (2017) p.9 in Lennox, J.C.: Cosmic Chemistry (2021) p.361.

34. Floridi, L.: *The Philosophy of Information.* Oxford Univ. Press (2011) p.131.

35. Sperry, R.W.: Psychology's Mentalist Paradigm and the Religion/Science Tension, *American Psychologist,* Aug (1988) p.609 in Lennox, J.C.: Cosmic Chemistry (2021) p.329.

36. Wald, G.: Life and Mind in the Universe, *International Journal of Quantum Chemistry* (1984) 26(S11):1 in Dirckx, S.: Am I just my brain. The Good Book Co. UK (2019) p.71.

37. See: Floridi, L.: *The Philosophy of Information.* Oxford Univ. Press (2011) p.79.
38. Pribram, K.: The neurophysiology of remembering, *Scientific American* 220 (1969) p.75.
39. See also the work of Wilder Penfield, Karl Lashley, Karl Pribram & Paul Pietsch from 1920s to 1970s in Wiles, R.B.: *The Mind in the Matrix – What the Complexity of the Universe tells us about Meaning.* Information Press Australasia (2019) p.104.
40. Floridi, L.: *The Philosophy of Information.* (2011) p.87.
41. Floridi, L.: *The Philosophy of Information.* (2011) p.40.

Chapter 7 – Classifying Information

1. Gibe, A.: Les Faux-monnayeurs [The Counterfeiters] 1925 in Encyclopaedia Britannica http://www.britannica.com
2. Lennox, J.C.: *Cosmic Chemistry: Do God and Science mix?* Lion Books, UK (2021) p.193.
3. Davies, P.: *The Fifth Miracle.* Penguin Books (1999) p.96.
4. Andrews, E.: *Who Made God? Searching for a Theory of Everything.* EP Books, England (2009) pp.150-152.
5. Lloyd, S.: *Programming the Universe.* Knopf, NY p.40 (2006) in Gregersen, N.H.: *'God, Matter and Information: towards a Stoicizing Logos Christology' in* Davies, P. & Gregersen, N.H. (eds) Information and the Nature of Reality. Cambridge Univ. Press (2010) p.334.
6. Bussell, J.: https://www.britannica.com/topic/cyberspace (Accessed 10/10/2020).
7. Popper, K.: *Epistemology Without a Knowing Subject.* Third International Congress for Logic, Methodology and Philosophy of Science, 25 August 1967. Amsterdam. (1968).
8. Penrose, R.: *Shadows of the Mind: A Search for the Missing Science of Consciousness.* Vintage Books (1995) pp.412-413.
9. Penfield, W.: *The Mystery of the Mind: A Critical Study of Consciousness and the Human Brain.* Princeton Univ. Press, NJ 76-78 (1975).
10. Pietsch, P.: *Shufflebrain: The Quest for the Holographic Mind.* Houghton Mifflin, Boston, MA (1981).
11. Interview of Bernard Carr by Robert Lawrence Kuhn: 'What Exists.' *On* ClosertoTruthTV. https://www.youtube.com/watch?v=79Tu6noBkWg – 21 Sep, (2023).
12. Meaning in this context implies that a reason or purpose exists.
13. Within the boundaries permitted within the Universe which some would term "limited will".
14. Interview of Sean Carroll by Robert Lawrence Kuhn: 'What Exists.' *On* Closerto TruthTV. https://www.youtube.com/watch?v=tYBeC4aDaKw – 25 Jun (2023).
15. McGrath, A.E.: *Natural Philosophy: On Retrieving a Lost Disciplinary Imaginary.* Oxford Univ. Press (2023) p.174.
16. Sir Roger Penrose & William Lane Craig: The Universe: How Did It Get Here and Why Are We Part of It?" Oct. 4, 2019, Brierly, J.: *The Big Conversation,* season 2, episode 2. https://www.youtube.com/watch?v=9wLtCqm72-Y
17. McGrath, A.E.: *Natural Philosophy: On Retrieving a Lost Disciplinary Imaginary,* Oxford Univ. Press, Oxford (2023) p.165.
18. Hawking, S. & Mlodinow, L.: *The Grand Design.* Bantam Press, London (2010) p.117-118.

Chapter 8 – Reconciling the Classical and Quantum Realms

1. Einstein, A.: 1933 lecture: often paraphrased: *"Make things as simple as possible, but no simpler."* See: Robinson, A.:'Did Einstein really say that?' Nature 557, 30 (2018). https://www.nature.com/articles/d41586-018-05004-4#:~:text=%E2%80%9CEverything%20should%20be%20made%20as,possible%20without%20having%20to%20surrender [Accessed 27.7.2024]
2. Wigner, E.:'The Unreasonable Effectiveness of Mathematics in the Natural Sciences.' *Communications in Pure and Applied Mathematics*, 13, No.1, Feb. (1960).
3. Shariatmadari, H.: *What is Reality?* Horizon, BBC Productions, UK (2011).
4. Currivan, J.: *The Cosmic Hologram – In-formation at the Center of Creation.* Inner Traditions, Rochester, Vermont (2017) p.106.
5. Glattfelder, J.B.: *Information - Consciousness – Reality: How a New Understanding of the Universe Can Help Answer Age-Old Questions of Existence.* Springer (2019) p.503.
6. Adams, D.: *The Hitchhiker's Guide to the Galaxy.* (1979)
7. Glattfelder, J.B.: *Information - Consciousness – Reality.* Springer (2019) p.496.
8. Hawking, S. & Penrose, R.: The Nature of Space and Time. Princeton Univ. Press, Princeton (1996) p.39 in Glattfelder, J.B.: *Information - Consciousness – Reality.* Springer (2019) p.492.
9. Glattfelder, J.B.: *Information - Consciousness – Reality.* Springer (2019) p.493.
10. Glattfelder, J.B.: *Information - Consciousness – Reality.* Springer (2019) p.492.
11. Heyerdahl, T.: The Kon-Tiki Expedition. (1950).
12. Al-Khalili, J.: *Black Holes, Wormholes & Time Machines.* Taylor & Francis, NY (1999) pp.106-7.
13. Al-Khalili, J.: *Black Holes, Wormholes & Time Machines.* Taylor & Francis, NY (1999) p.199.
14. Liu, B. & Bromm, V.: Accelerating Early Massive Galaxy Formation with Primordial Black Holes. *The Astrophysical Journal Letters.* 937 (2): L30. arXiv:2208.13178 (2022).
15. Turok, N.: There could be a 'mirror image' of our universe on the other side of the big bang. *in* Cossins, D.: (ed) How to Think about Reality, *New Scientist*, England (2025) p.63.
16. For further discussion of white holes see Wiles, R.B.: *The Information Paradox* Information Press (2014) p.264.
17. Genesis 1:1 (NIV) Those who doubt the veracity of the Bible as a credible source of information might like to look at its internal consistency and the accuracy of its predictions, such as God's letter to Cyrus the Great (c.600-c.530 BCE) recorded in Isiah Chapter 45 (written c.700 BCE), that describes Cyrus' exploits 100 years before he was born or the writings of Daniel Chapters 7 & 11 (written c.530 BCE) that describe the rise of Alexander the Great (356-323 BCE) & his successors two to four centuries later; then assemble the evidence for a multiverse and compare that evidence with the Bible.
18. Al-Khalili, J.: *Paradox. The Nine Greatest Enigmas in Physics.* Black Swan, UK (2013) pp.250-254.
19. Hawking, S. & Mlodinow, L.: *The Grand Design* (2010) p.63.
20. Al-Khalili, J.: *Paradox: The Nine Greatest Enigmas in Physics.* Black Swan, UK (2013) pp.252-253.
21. Yazgin, E.: Quantum Simulations. *Cosmos* 105 (2024) p.38.
22. McGrath, A.E.: *Natural Philosophy.* Oxford Univ. Press, Oxford (2023) p.162.
23. Lewton, T.:'Inside the event horizon.' *New Scientist* 15 Jun (2024) pp.32-35.
24. For further discussion of holograms see: Wiles, R.B.: *The Information Paradox.* Information Press (2014) pp.254-256.

25. For further discussion of the Holographic Universe see: Bekenstein, J.D.: 'Information in the Holographic Universe.' *Scientific American*, Aug (2003) & Glattfelder, J.B.: *Information - Consciousness – Reality*. Springer (2019) pp.496-500.

26. Al-Khalili, J.: *Paradox: The Nine Greatest Enigmas in Physics*. Black Swan, UK (2013) pp.227-229.

27. For more background on David Bohm's understanding of quantum mechanics see: Wiles, R.B.: *The Information Paradox*. Information Press (2014) pp.255-259.

28. Cossins, D.: (ed) How to Think about Reality. *New Scientist* England (2025) p.43.

29. Batiz, Z. & Chauhan, B.C.: *Holographic Principle and Quantum Physics*. 10 Nov 2006 arXiv:physics/0611104 p.17.

30. Hoffman, D.: 'We assume our senses see reality as it is, but that could be just an illusion of evolution.' in Cossins, D.: (ed) How to Think about Reality. *New Scientist* England (2025) p.81.

31. Strassler, M.: 'The language we use in physics obscures some of our most beautiful and fascinating discoveries.' *in* Cossins, D.: (ed) How to Think about Reality. *New Scientist* England (2025) p.51.

32. Huggett, N. & Rovelli, C.: Quantum Spacetime *in Scientific American* Special Edition Vol. 34:2 (2025) p.12.

33. Batiz, Z. & Chauhan, B.C.: *Holographic Principle and Quantum Physics*. 10 Nov 2006 arXiv:physics/0611104 p.4.

34. Pratt, D.: David Bohm and the Implicate Order. *Sunrise Magazine* Theosophical Univ. Press Feb/Mar 1993.

35. McGrath, A.E.: *Natural Philosophy*. Oxford Univ. Press, Oxford (2023) p.116.

Chapter 9 – Constructing the Cosmos from Information

1. Bekenstein, J.D.: 'Information in the Holographic Universe.' *Scientific American*, Aug 2003 p.65.

2. Denton, M.: 'Evolution: A Theory in Crisis Revisited.' *Inference*, Part 3 of 3. Vol. 1 No.3 July 2015.

3. Lennox, J.C.: *Cosmic Chemistry: Do God and Science mix?* Lion Books, UK (2021) p.340.

4. Lennox, J.C.: *Cosmic Chemistry*. Lion Books, UK (2021) p.337.

5. Davies, P.: *The Cosmic Blueprint*. Simon and Shuster, NY (1988) p.203.

6. Rees, M.: *Just Six Numbers: The Deep Forces that Shape the Universe*. Weidenfield & Nicholson, London (1999).

7. Bradley, W.L.: 'Nature. Design or Designoid' in Dembski, W.A.: (ed)*Mere Creation. Science, Faith and Intelligent Design*. InterVarsity Press (1998) p.40.

8. Davies, P.: 'Universe from bit.' *in* Davies, P. & Gregersen, N.H.: (eds) *Information and the Nature of Reality*. Cambridge Univ. Press (2010) p.75.

9. Lennox, J.C.: *Cosmic Chemistry*. Lion Books, UK (2021) p.326.

10. Lennox, J.C.: *Cosmic Chemistry*. Lion Books, UK (2021) p.345.

11. Davies, P.: *Physics and the Mind of God: The Templeton Prize Address* (1995) https://www.firstthings.com/article/1995/08/physics-and-the-mind-of-god-the-templeton-prize-address.

12. Laughlin, R.: *A Different Universe: Reinvinting Physics from the Bottom Down*. Basic Books, NY (2005) pp.168-169.

13. Lennox, J.C.: *Cosmic Chemistry*. Lion Books, UK (2021) p.201.

14. Hoyle, F.: *Evolution from Space*, Omni Lecture, Royal Institution, London, 12 January 1982. University College Cardiff Press, 1982. Hoyle was not the first to propose panspermia but he was one of its most high-profile proponents.

15. Dawkins, R.: *The Greatest Show on Earth: The Evidence for Evolution*. Bantam Press, Great Britain (2009) p.416.

16. Davies, P: *The Fifth Miracle: The Search for the origin of Life*. Penguin Books (1999) p.86.

17. Huda, A. & Bushel, P.R.: *Widespread Exonization of Transposable Elements in Human Coding Sequences is Associated with Epigenetic Regulation of Transcription*. Published online 2013 Jun 19. See ENCODE at The NIH National Human Genome Research Institute. www.genome.gov

18. Ribak, E.: 'The Purpose of Our Eyes' Strange Wiring Is Unveiled.' *Scientific American*, 18 Mar 2015 https://www.scientificamerican.com/article/the-purpose-of-our-eyes-strange-wiring-is-unveiled/ - Reprinted from *The Conversation*, UK 13 Mar (2015).

19. Baylor, D.A., Lamb, T.D. and Yau, K.W.: Responses of retinal rods to single photons. *Journal of Physiology*, Lond. 288, pp.613-634 (1979).

20. For more examples and further discussion of this topic see: Wiles, R.B.: *The Information Paradox*. Information Press (2014) pp.181-187.

21. Marks, R.J. II, Dembski, W.A. & Ewart, W.: *Introduction to Evolutionary Informatics*. World Scientific Publishing Co. Singapore (2017) p.98.

22. For further discussion regarding bacterial resistance to antibiotics, see: Wiles, R.B.: *The Information Paradox*. Information Press (2014) pp. 223-228.

23. Boyle, A.: *The not-so-angry evolutionist*. 15 Oct (2009). https://www.nbcnews.com/science/cosmic-log/not-so-angry-evolutionist-flna6c10404299

24. Lennox, J.C.: *Cosmic Chemistry*. Lion Books, UK (2021) p.328.

25. Gould, S. J.: *The Structure of Evolutionary Theory*. (2002) p.749.

26. For further discussion of Punctuated Equilibrium, see: Wiles, R.B.: *The Information Paradox*. Information Press (2014) pp.166-168.

27. For further discussion regarding the information generators, see: Wiles, R.B.: *The Information Paradox*. Information Press (2014) pp.284-6 & 357-9.

28. Küppers, B.O.: in Lennox, J.C.: *Cosmic Chemistry*. Lion Books, UK (2021) p.205.

29. Lennox, J.C.: *Cosmic Chemistry*. Lion Books, UK (2021) p.199.

30. Lennox, J.C.: *Cosmic Chemistry*. Lion Books, UK (2021) p.206.

31. Lennox, J.C.: *Cosmic Chemistry*. Lion Books, UK (2021) p.200.

32. Davis, P. & Kenyon, D.H.: *Of Pandas and People: The Central Question of Biological Origins*. Haughton Pub. Co., Dallas, Texas (1989) p.7.

33. Lennox, J.C.: *Cosmic Chemistry*. Lion Books, UK (2021) p.334.

34. For further discussion of Design see: Wiles, R.B.: *The Mind in the Matrix*. Information Press Australasia pp.116-118.

35. Lennox, J.C.: *Cosmic Chemistry*. Lion Books, UK (2021) p.367.

36. Wigner, E.P.: 'The unreasonable effectiveness of mathematics.' *Communications in Pure and Applied Mathematics*, 13 (1960) pp.1-14; see Lennox, J.C.: *God's Undertaker: Has Science Buried God?* Lion Books, Oxford (2007) p.60.

37. For further discussion of the complexity of life see: Wiles, R.B.: *The Information Paradox*. Information Press (2014) pp.126-127.

38. For further discussion regarding fine-tuning see: Wiles, R.B.: *The Information Paradox*. Information Press (2014) pp.201-204.

39. For further discussion regarding biological efficiency see: Wiles, R.B.: *The Information Paradox*. Information Press (2014): ATP Synthase p.94; citric acid cycle p.113; histones p.106; reproduction p.177-8 & embryogenesis pp.124-6.

40. Davies, P.: *The Fifth Miracle*. Penguin Books (1999) p.39.

41. Davies, P.: *The Fifth Miracle*. Penguin Books (1999) pp.40-41.

42. Whitworth, B.: *Quantum Realism*. Chapter 2, p.18 (2025). https://brianwhitworth.com/quantum-realism/

43. Cossins, D.: (ed) 'Will we ever have a theory of everything?' in How to Think about Reality, *New Scientist*, England (2025) p.88.

44. Cossins, D.: (ed) 'Will we ever have a theory of everything?' in How to Think about Reality, *New Scientist*, England (2025) pp.86-87.

Chapter 10 – Information Mechanics: A New Understanding

1. Hoyle, F.: in Stockwood, M.: (ed) *Religion and the Scientists: Addresses Delivered in the University Church, Cambridge, Lent Term*, 1957, SCM Press: London (1959) p.64.

2. Davies, P.W.C.: *The 5th Miracle*. Penguin Books (1999) pp.32-33.

3. Lennox, J.C.: *Cosmic Chemistry*. Lion Books, UK (2021) p.337.

4. Shapiro, J.: Interview with Mazur, *The Paradigm Shifters*, p.23 *in* Lennox, J.C.: Cosmic Chemistry. Lion Books, UK (2021) p.306.

5. Behe, M.J.: Darwin's Black Box: *The Biochemical Challenge to Evolution*. Free Press, NY (2006) pp.39.

6. Sacks, J.: *The Great Partnership*. Hodder & Stoughton, London (2011) pp.2-3.

7. Lennox, J.C.: *Cosmic Chemistry*. Lion Books, UK (2021) p.329.

8. Davies, P.: 'Universe from Bit.' *in* Davies, P. & Gregersen, N.: (eds) *Information and the Nature of Reality: from Physics to Metaphysics*. Cambridge Univ. Press (2010) p.75.

9. Chalmers, D.J.: *Explaining Consciousness Scientifically – Choices and Challenges*. The Science of Consciousness Conference, Tucon, Arizona 1994 https://www.youtube.com/watch?v=_lWp-6hH_6g

10. Andrews, E.H.: *Who Made God? Searching for a Theory of Everything*. EP Books, Darlington, UK (2009) p.148.

11. Andrews, E.H.: *Who Made God?* EP Books, Darlington, UK (2009) p.152.

12. Noble, R. & Noble, D.: Was the Watchmaker Blind? Or Was She One-Eyed? Biology 6(4):47 (2017) *in* Lennox, J.C.: *Cosmic Chemistry*. Lion Books, UK (2021) p.312.

13. Sacks, J.: *The Great Partnership*. Hodder & Stoughton, London (2011) p.9.

14. See the Torah – Genesis 1:1 and the Bible – John 1:1-3.

15. Lennox, J.C.: *Cosmic Chemistry*. Lion Books, UK (2021) p.365.

16. Dawkins, R.: *The Greatest Show on Earth*. Bantam Press, London (2009) p.220.

17. Lennox, J.C.: *Cosmic Chemistry*. Lion Books, UK (2021) p.326.

18. Lennox, J.C.: *Cosmic Chemistry*. Lion Books, UK (2021) p.328.

19. Lennox, J.C.: *Cosmic Chemistry*. Lion Books, UK (2021) p.179-180.

20. For further discussion of the Information Model see: Wiles, R.B.: *The Mind in the Matrix*. Information Press Australasia p.106-108.

21. Walker, S.I., Davies, P.C.W. & Ellis, G.F.R.: *From Matter to Life: Information and Causality.* Cambridge Univ. Press (2017) p.1.

22. Chaitin, G.J.: *The Unknowable.* Springer, NY (1999) p.106 *in* Lennox, J.C.: *Cosmic Chemistry.* Lion Books, UK (2021) p.332.

23. Sagan, C.: *Pale Blue Dot: A Vision of the Human Future in Space.* Random House, NY (1994).

24. Whitworth, B.: Quantum Realism. https://brianwhitworth.com/quantum-realism/ (2025) Ch 1, p.5.

25. Whitworth, B.: Quantum Realism. https://brianwhitworth.com/quantum-realism/ (2025) Ch 2, p.17.

26. *Pilot Wave Theory and Quantum Realism.* PBS Digital Studios (2016). https://www.youtube.com/watch?v=RlXdsyctD50 [Accessed: 12 Jun 2025]

27. Lennox, J.C.: *Cosmic Chemistry.* Lion Books, UK (2021) p.333.

28. Lennox, J.C.: *Cosmic Chemistry.* Lion Books, UK (2021) p.361.

29. Wiles, R.B.: *The Mind in the Matrix.* Information Press, Australasia p.116.

30. Lennox, J.C.: *Cosmic Chemistry.* Lion Books, UK (2021) p.365.

31. Sacks, J.: *The Great Partnership.* Hodder & Stoughton, London (2011) p.9.

32. Lennox, J.C.: *Cosmic Chemistry.* Lion Books, UK (2021) p.371.

33. Wiles, R.B.: *The Information Paradox.* Information Press (2014) p.335.

34. Whewell, W.: *The Philosophy of the Inductive Sciences*, John W. Parker, London (1847) Vol.2, p.36 *in* McGrath, A.E.: *Surprised by Meaning: Science, Faith and How We Make Sense of Things.* Westminster John Knox Press, Kentucky (2011) p.2.

35. This statement is similar to an observation made by Francis Bacon, the Father of the Scientific Method, who wrote in 1625: "*A little philosophy inclineth man's mind to atheism, but depth in philosophy bringeth men's minds about to religion.*" Essays: *Of theism* (1625) *in* Ratcliffe, S.: Oxford Essential Quotations (4th ed.) Oxford Univ. Press (2016). The intervening three centuries have not changed the concept.

INDEX

Aastrup, Johannes ... 117
Adami, Christoph ... 61
Agency ... 121-123, 131
AI (See Intelligence)
Alberts, Bruce ... 22
Algorithmic Information Theory (AIT) ... 79, 127
Al-Khalili, Jim ... 91-2, 94, 98
Alzheimer's Dementia ... 73-74
Amoeba ... 69, 74, 77, 96, 123
Amplituhedron ... 118
Anaesthetic agents ... 70-71
Ancient Greeks ... 18
Anderson, Carl ... 130
Andrews, Edgar ... 52, 122
Anti-matter ... 46, 87
Appendix ... 112
Aquinas ... 131
Aristotle ... 131
Armstrong, David ... 49
Asimov, Isaac ... 127
ATP-synthase generator ... 115
Awareness (see Consciousness)

Babbage, Charles ... 19
Babcock, Nathan ... 70
Babylonian tablets ... 14
Bacteria ... 34, 56, 61, 72, 108, 112
  Chromophore resonance ... 73
Batiz, Zoltan ... 100
Beauregard, Mario ... 57, 74
Behe, Michael ... 58, 62, 121
Bekenstein, Jacob ... 20, 37, 107
Big Bang ... 92, 109, 116, 124, 126-127
Biogenesis ... 64
Black body radiation ... 46
Black holes ... 51, 90-92
Bohm, David ... 93, 98, 102, 104
Bohr, Niels ... 93
Bottom-up analysis ... 59, 61,120-121, 123-125
Brain ... 15, 25, 34, 38-39, 50, 52, 65-66, 68-70, 73-78, 83, 95-96, 109, 111, 121
  Electro-magnetic field ... 73
  Models ... 69
  Oscillating ... 73
  Quantum ... 68-70, 73, 75-6
Braunstein, Samuel ... 51
Brennan, Gavin ... 94
Buckyballs ... 94

Cambrian Explosion ... 109-110, 113-114, 120, 125
Carbon ... 47-48
Carr, Bernard ... 83
Carroll, Sean ... 85, 87
Causal power ... 39, 41, 127, 131
Causation ... 42, 75, 127
  Bottom-up ... 64
  Top-down ... 57, 64, 130
Chaitin, Gregory ... 79, 127
Chalmers, David ... 22, 68, 121
Chauhan, Bhag C. ... 100
Choice ... 43, 53, 83, 92, 117, 121-124
Church, Alonzo ... 19
Citric acid cycle ... 115
Clarke, Stuart ... 48
Classical Mechanics ... 18, 22-24, 29, 31, 53, 75, 77, 89, 124
Clockwork (see Universe)
Close, Frank ... 46
CNET ... 69
Complexity ... 47, 50, 55, 60-63, 79, 104, 112-114
  Boundary ... 58
  Counterflow of ... 119-120
  Irreducible ... 58-59, 121, 125
  Measurement of ... 79
Computer(s) ... 13, 15, 20, 24, 33-34, 38-39, 43, 51-52, 62, 66, 70, 76-78, 80-83, 111, 118, 120-122, 131
  Quantum ... 68
Concussion ... 74
Conscious EM Field Information Theory ... 73
Consciousness ... 15, 22,24, 32-34, 39, 47, 50, 52-53, 56, 66, 68-74, 76-77, 83, 86-87, 96, 105, 108-110, 112, 115, 121-122, 126, 130
Context ... 33-34, 37-38, 40, 42, 55
Coyne, Jerry ... 58
Craig, William Lane ... 30
Cyberspace ... 13, 81-83, 85-87, 100-101, 108, 115-116
  Map ... 82
Cytoskeleton ... 69, 77, 95-96

Dark energy ... 30
Dark matter ... 30
Darwin, Charles ... 57, 61, 111
  Tree of Life ... 108
Data ... 13, 18, 32-34, 38, 40, 42, 59, 73, 77-78, 85
Data Grounding Problem ... 40
Davies, Paul ... 14, 16, 20, 31, 39, 43, 45, 47-49, 55-57, 62, 80, 110-112, 115
Dawkins, Richard ... 52, 61, 111-112, 125
de Broglie, Louis ... 93
de Duve, Christian ... 22
Decoherence ... 72, 74, 94-95, 103
Democritus ... 18, 24

Dennett, Daniel....69
Denton, Michael....108
Design....49, 52, 62, 112, 114, 119, 128, 131
Determinism....18, 43, 53, 75, 80, 83, 105, 114, 124
Digital Philosophy....19
Digital Physics....19
Dirac, Paul....46, 130
Dirckx, Sharon....29
DNA....13, 24, 33-34, 41, 49, 59, 113-114, 116, 125-126
  code....22, 24, 32, 93, 108, 128-129
Dualism....77-78
Dunbar, Robin....67
Dyson, Freeman....48

Einstein, Albert....18, 22, 27, 30, 43, 87, 102-103, 130
Eldridge, Niles....108, 113
Electromagnetic field (see fields)
Electromagnetic Force Constant....47
Ellis, George....16, 39, 55, 126
Embryogenesis....115
Emergence....25, 52, 117, 121, 126, 129
Emotions....96
ENCODE Project....111
Engelhardt, Netta....96
Entanglement....68, 70, 73, 93, 108
Entropy....33, 37, 90, 115-116, 119-120
  Law of....15, 22, 47, 119, 125
Enzymes....60
Epigenome....58-59, 64, 109, 112, 114,122
Euclidian space....85
Evolution....56, 59, 61, 64, 111-113, 122-123, 128-129
  Darwinian....61-64, 109-110, 114
  Macro....112, 125
  Micro....112
Explanatory gap....68

Feelings....66, 85-86, 96
Fermi, Enrico....130
Fermions....117-118
Fields....21, 43, 98-99, 107
  Electromagnetic....70-71, 73
  Gravitational....115
  Information....99
  Quantum....70, 73, 98-99, 117-118
  Zero-point....84
Fine tuning....22, 47, 49, 114, 125
First organism....62, 128-129
Floridi, Luciano....17, 19, 22, 28, 31, 34, 38, 40-41, 51, 75, 78
Fluorescence....70-71
Forces....29, 48, 98, 122
  Electromagnetic....47
  Fundamental....47, 122
Fredkin, Edward....19
Free will....83
Fröhlich coherence....71, 73
Fröhlich, Herbert....71
Fundamental entities....28-29, 80

Gabor, Dennis....97
Galileo....45
Gaps (see Information Gaps)
Genes/Genome....23, 41, 59, 61, 111, 115
Glattfelder, James....51, 90
Gleick, James....21, 46
God....41, 92-93, 98, 111, 127, 132
  Mind of....123, 130
God of the gaps (see Information Gaps)
Goldilocks Zone....114
Gould, Stephen Jay....58-59, 108, 113
Gravitation wave....130
Gravitational Force Constant....47
Gribbin, John....43
Grimstrup, Jesper Moller....117

Hameroff, Stuart....68-69, 71-72, 95, 122
Hardware....13, 25, 38-40, 52, 56, 58, 60, 62, 66, 76-77, 80, 82, 96, 121-122
Hartle, James....30
Hawking, Stephen....29-30, 87, 91
Heisenberg, Werner....132
Heyerdahl, Thor....91
Higgs Boson....129
Hilbert space....85
Hoffman, Donald....71, 100
Holograms....70, 74, 76, 97
Holographic Brain Theory....70
Hoyle, Fred....48, 111, 119, 124
Huggett, Nick....103
Human Genome Project....23, 112

Immateriality (see Non-materiality)
In-determinism....53,124
Infinite Monkey Theorem....61
Information....28, 31, 38, 42
  Age....13
  Algorithmic....32, 63, 79-80, 127
  Analogue....46
  Biological....38-39, 58, 63, 115
  Bundling....41
  Characteristics....38
  Code....109
  Compositional....32, 38
  Compressible....38, 50, 79-80
  Conservation of....51
  Content....21, 28, 34, 37, 40-41, 52, 64, 79-80, 90, 102
  Definition....31

Descriptions of 79, 97
Digital 46
Discontinuities 93, 108, 110, 113, 124, 130
Domains 13-14, 80, 85-87, 89-90, 101, 104, 116
Downloads 73, 108, 110, 113, 123-124
Fabric 84, 99, 102-103, 124
False 67
Functions of 37, 39
Fundamental entity 14, 21,28, 31, 46, 51, 81, 118, 130
Gaps 108-109, 110-111
 Aliens of the 111
 Evolution of the 111
 God of the 111
 Materialism of the 125
 Naturalism of the 111
 Of ignorance 109-111, 113
 Science of the 110
 Structural 110-111, 113
Generation 50, 52, 62, 108, 113
Golden Rule of 51
Incompressible 38, 80, 113
Initial influx 52, 117, 123-125
Integration of 120
Life, in 58
Meaningful 20, 22-24, 32-35, 37-41, 50-51, 55, 62-63, 78-80, 83, 85, 108, 113, 115-6, 125, 131
Measuring 20, 32-33, 79, 95, 115, 119
Mechanics 16, 19, 23, 53, 64, 119, 126-127, 131
Model 117, 119, 126, 128, 132
 Testing 126
Mysteries 109
Philosophy of 19
Pragmatic 32, 38
Qualitative 32, 37-38, 79
Quantitative 37-38, 79
Semantic 22-23, 32, 37-38, 78-79, 85, 116, 121
Shannon 20, 23, 32, 37-38, 79-80, 85, 119
Stochastic 38, 87
Structured 41, 45
Syntactic 20, 32, 37, 85
Time, and 27, 41-42, 61-63, 68, 81, 84-85, 92, 97
Universal Rule of 51
Infosphere 38, 107, 116-117, 124, 127
 Map 117
Initial conditions 50, 91-92, 115, 117, 123-124, 127-128
Intelligence
 Artificial 24, 33-34, 131
 Natural 24, 34

Jibu, Mari 69
Junk-DNA 111-112

Kantor, Frederick 19
Kaufman, Stuart 61
Kenyon, Dean 114
Kepler, Johannes 45
Kerskens, Christian 68-69
Khan, Sana 70
Knowledge 13, 17, 40-42, 55, 67, 76-77, 109-110, 114, 127
Kolmogorov, Andrey 79
Krakauer, David 29
Kuhn, Thomas 23
Küppers, Bernd-Olaf 50, 58, 62-63, 65, 113

L-Functions & Modular Forms Data Base 89
LaBar, Thomas 61
Landauer, Rolf 49
Lasers 70, 84, 97, 104
Lashley, Karl 76, 95
Laughlin, Robert 111
Laws of Physics 14, 20, 50
Law of Thermodynamics, (2nd) (see Entropy)
Lennox, John 18, 30, 40-41, 49, 51-52, 58, 60, 63-64, 75-76, 110-111, 114, 121-125, 127, 130
Leucippus 18
Levine, Joseph 68
Life 15, 22, 47-49, 55-64, 72, 77, 80-81, 83, 105, 108-109, 112, 114, 122, 125, 128
 Complexity in 47, 56, 59, 109, 114, 129
 Definition of 55-57
 Finely tuned 47, 110
 Origin of 56, 60, 62-64, 93, 108, 110, 112, 120, 124, 130
Lloyd, Seth 20, 27, 50
Logic 16, 28, 43, 53, 84, 107, 124, 127, 130
Loop Quantum Gravity 24
Lovelace, Ada 19
Lowe, Ian 13

M-Theory 24, 87, 130
Machine language 33, 38-39
Machine learning 33
MacKay, Donald 31
Magnetic dipole oscillations 71
Many Worlds (see Quantum Interpretations)
Marks, Robert J.II 33
Materialism 15, 18, 24, 29, 39, 74, 81, 83, 114, 125-126, 131
Mathematical Theory of Communication (MTC) 20, 37-38, 79
Mathematics 14, 18, 32, 37, 45-46, 51, 81, 89-91, 97, 114, 117, 128, 130

Matter waves....93, 102
Matter/energy....24, 28, 30-31, 81, 127
Mattergram....97, 102
McFadden, John....73
McGrath, Alistair....16, 86, 96, 104
McLuhan, Marshall....21
Meaning....20-22, 32, 38-41, 50, 61-63, 65, 68, 78-80, 83, 85-86, 120-121, 123, 127-128
  Interpretation of....78
Mechanisms....46, 53, 57-59, 62-63, 69, 71, 73-74, 77, 93, 95-96, 110, 113, 120-125, 127, 131
Memory....34
  Holographic....70, 76, 82, 104
  Long-term....74, 104
  Short-term....68, 73, 104
Meta-information....58, 60, 76, 114, 121-122, 131
Microtubules....69-70, 72-73, 77
Mind(s)....24, 28, 33, 39, 63, 65-66, 68, 73, 77, 126
  Philosophy of....67
  Quantum....68, 74, 104
  Significance of....33
  Theory of....67
Mind-brain Complex....39, 68, 71, 74, 76, 77, 83, 104
Mindspace....66, 68, 82-87, 95-96, 104, 108, 116, 132
  Map....83
Misnomers....112
Mitchell, John....90
Mitochondria....60
Mlodenow, Leonard....29, 87
Models....24, 28-30, 33, 41, 53, 59, 62, 69, 73-74, 76, 79, 95, 101, 109, 129
  Orch-OR....69
  Universe....29, 48, 119, 126-128
Modern synthesis (see Evolution)
Molnar, Tibor....24, 28, 30-31, 33
Monism....77-78
Morality....112
Müller, Gerd....62
Multiverse....92
Mysteries....18, 46, 48, 68, 81, 109

Nagle, Thomas....74
Neo-Darwinian Synthesis (see Evolution)
Neurology....76-77
Neuroscience....76-77
Newton, Isaac....42, 45, 89
Nishiyama, Akihiro....69
Noble, Dennis....59-60, 122
No-hiding Theorem....51, 93, 102
Non-materiality....13, 21, 25, 28-29, 32-33, 39-40, 42, 49-50, 56-57, 65, 68, 75-78, 80-82, 84-85, 126, 128
Nyquist, Harry....19

Oscillations....70-74
Overman, Dean....63

Panspermia....111
Paradigms....22, 109
  Current....107, 128
  Information....125, 129
  Material....21, 125, 128
Paradigm shift....23, 126
Paradox(es)....15
Particle(s)....3, 21, 43, 46, 65, 92, 94, 98-99, 118, 122, 129
  Fundamental...21, 80-81, 93, 101, 103, 118
  Phantom....30, 52, 80
  Physics....18, 118
  Sub-atomic....13, 18, 30, 84, 109
  Virtual....30, 98
  Wave duality....43, 85, 93, 101
Pati, Arun....51
Pauli, Wolfgang....130
Pearcey, Nancy....40, 50
Penrose, Roger....69, 72, 81-82, 86, 91, 95, 118
Phantom particles (see Particles)
Philosophy....29, 42, 76-77, 105, 109, 123
  Digital....19
  Of information....19
  Of mind....67
  Natural....87
Phosphorescence....71
Pilot-wave Theory....93, 128
Planck, Max....18, 46, 130
Planck scale....13, 28, 58, 100-101
  Length....28, 95
  Time....101
Plants....34, 69, 71-72, 110, 123-124
Plasma....104
Plato....45, 77
Popper, Karl....81, 86, 126
  Three-World Theory....81, 85
Positron....46, 130
Pre-conditions (see Initial Conditions)
Presuppositions....127-128
Pribram, Karl....76, 95
Protein (molecular) machinery....58, 63
Psychiatry....76-77
Psychology....76-77
Punctuated Equilibrium, Theory of....113
Pythagoras....45

Quantisation....31, 46, 110

Quantum....43
Beats....73
Biology....70, 77
Brain Dynamics (QBD)....69
Coherence....71
Domain....31, 53, 58, 66, 68, 85, 90, 93, 97, 131
Domain map....84
Electric effect....71
Fabric....116
Fields....70, 73, 98-99, 117-118
Gravity....103
Information....13, 58, 77, 84, 96, 101, 116, 127
Leakage....94, 105
Mechanics....15, 18, 22-24, 28, 31, 43, 51, 53, 85, 90, 92-94, 97-98, 100, 114, 120, 122, 124, 126, 128
Mind....68, 74, 77, 104
Oscillations (see Oscillations)
Realism....127-129
Realm (see Quantum domain)
Vacuum....30-31, 84
World (see Quantum domain)
Quantum interpretations
Copenhagen....93, 128-129
de Broglie-Bohm....93, 128-129
Many Worlds....92, 129
Pilot wave....93, 128

Radioactive decay....98, 102
Reason(ing)....34, 42, 115, 127
Reductionism....16, 59-60, 64, 74, 86, 120-121, 123
Redundancy....69, 114
Rees, Martin....22, 47
Relativity Theory....15, 18, 27, 53, 117, 124
General....46, 90-91, 100, 117
Special....81
Ribosomes....60
RNA....110
RNA, messenger....60, 114
Rovelli, Carlo....27-28, 46, 103

Sachs, Oliver....66
Sacks, Jonathan....121, 123, 127
Schrödinger, Erwin....18
Schrödinger's cat....94
Equation....102
Schwartz, Jeffrey M.....57, 74
Schwarzschild, Karl....90
Searle, John....49, 52
Self....76-77
Self-awareness....22, 34, 47, 66, 76-77, 109, 115
Self-organisation....61-62, 113, 128
Self-regulation/determination....57, 109
Seneca the Younger....65
SETI: Search for Extra-Terrestrial Intelligence....111
Shannon, Claude....19-20
Shannon Theory....20, 32, 37, 119-120
Shapiro, James....120
Shepphard, R.Z.....38
Singularity....91-92, 110, 124
Sleep....74
Software....13, 25, 38-40, 55-56, 66, 77, 82, 123, 131
Downloads (see Information)
Solar System....47, 55, 125
Space-time....13, 21, 23-24, 27-30, 32, 39, 53, 66, 69, 80-81, 85-86, 127
Domain....80, 93-95, 108, 123
Fabric....30, 32, 80-81, 84, 90, 94, 100, 102, 104, 107, 109, 116, 122, 125
Map....80
Pixelation....95, 100
Sperry, Roger....75
Spin transfer....71
Spinors....117-118
Standard Model of Cosmology....24, 48, 92-93, 107, 114, 127-128
Standard Model of Physics....15, 30, 110, 126
Stapp, Henry....57, 74-75, 77, 130
Stenger, Viktor....69
Strassler, Matt....101
String Theory....24
Sub-radiance....71
Superposition....73, 94, 99
Super-radiance....70-71
Susskind, Leonard....89
Synchronous oscillations (see Oscillations)
Systems....27, 32, 53, 58-59, 74, 84, 124
Biology....59-60
Complex....59, 109, 121
Data....34
Hardware-based....62, 66
Information....23, 32, 34, 56, 59, 114, 123
Analogue....57, 60
Digital....57, 60
Meta....32
Irreducibly complex....58-59, 121, 125
Living....58-60, 109, 112
Physical....27, 74
Quantum....95, 103
Self-organising....61-62
Simple....59, 61
Software-based....62
Top-down....121, 124

Tau protein....73
Tegmark, Max....72, 90
Telegraphers....19

Terahertz ... 71
Testing Theory ... 41
Theology ... 76-77
Thought ... 24, 49-50, 60, 65-66, 69, 73, 75-77, 81-83, 85, 96, 101, 108-109, 121, 131
Time ... 27, 41-42, 61, 68, 81, 85, 92, 97, 100
Timpson, C.G. ... 49
Top-down analysis ... 57, 59-60, 64, 76, 121, 123-125, 130
Transcranial Magnetic Stimulation (TMS) ... 71
Trilobites ... 110
Trnka, Jaroslav ... 118
Truth ... 28, 41
Tubulin ... 70, 74
Tunnelling ... 99
Turing, Alan ... 19
Turing Machine ... 51
Turok, Neil ... 92
Twistor Theory ... 118

Understanding ... 22, 34, 42, 50, 76-79, 127, 129
Universe
- Clockwork ... 18, 42
- Dimensions ... 87, 130
- Formation of ... 29-30, 52, 92, 107, 109-110, 120, 124-125, 129
- Front-loading ... 110, 130
- Fundamental entities ... 29, 46, 68
- Fundamental forces ... 47
- Holographic Theory ... 96-97, 99
- Information-first ... 52, 130
- Laws of ... 34, 45-46, 63, 65, 82, 108, 110, 115-116, 124-126, 129
- Map ... 117
- Materialistic ... 18, 39, 43, 80-81, 83, 87, 94, 117, 119
- Mathematical ... 45, 89-90
- Natural Laws ... 39, 55, 82, 108-109, 121-123
- Word-based ... 41

Unmoved mover ... 131

Viruses ... 56
von Beyer, Hans Christian ... 31
von Neumann, John ... 19

Wald, George ... 76
Walker, Sara Imari ... 16, 39, 55-57, 59, 126
Wallace, Alfred Russell ... 110
Water ... 47-48
Wavicles ... 101
Weaver, Warren ... 20
Wet-ware ... 76
Wheeler, John Archibald ... 3, 19, 21, 31, 90
Whewell, William ... 13, 131
White holes ... 91-92
Whitworth, Brian ... 72, 116, 127
Wiener, Norbert ... 19, 37
Wigner, Eugene ... 22, 69, 89
Williams, George ... 49
Woit, Peter ... 117
Wisdom ... 13, 42, 76-77
Worldview ... 22, 75

Yasue, Kunio ... 69
Yockey, Hubert ... 63

Zeilinger, Anton ... 31
Zeitgeist ... 22
Zero-point field ... 84
Zuse, Konrad ... 19

ABOUT THE AUTHOR

Robert Wiles is a senior lecturer in the Rural Clinical School at the Australian National University and is the rural academic co-ordinator for the Snowy Mountains region south of Canberra, a role he has had for the past 20 years. After graduating from the Faculty of Medicine, University of Sydney, he spent five years in the United Kingdom pursuing post-graduate studies in obstetrics, anaesthetics, orthopaedics and general practice before returning to Australia. He has been a rural generalist in Cooma for over 30 years during which time he has, for varying periods, been a visiting medical officer in emergency medicine, anaesthetics, obstetrics and in-patient care at the Cooma District Hospital in conjunction with general practice. He has been a medical educator in undergraduate and postgraduate rural medicine throughout this period.

He has been married for over 35 years with four adult children and two grandchildren and has been involved with Christian schooling for over 25 years.

He has a passion for teaching and learning in Medicine and Science which, over the past 20 years has extended into exploring and understanding the information that defines our universe. His previous books are:

The Information Paradox, published in 2014

The Mind in the Matrix, published in 2019

www.ingramcontent.com/pod-product-compliance
Lightning Source LLC
LaVergne TN
LVHW050915210926
848482LV00012B/443
* 9 7 8 0 9 8 7 5 6 2 2 3 4 *